CAMBRIDGE LIBRARY COLLECTION

Books of enduring scholarly value

Earth Sciences

In the nineteenth century, geology emerged as a distinct academic discipline. It pointed the way towards the theory of evolution, as scientists including Gideon Mantell, Adam Sedgwick, Charles Lyell and Roderick Murchison began to use the evidence of minerals, rock formations and fossils to demonstrate that the earth was older by millions of years than the conventional, Bible-based wisdom had supposed. They argued convincingly that the climate, flora and fauna of the distant past could be deduced from geological evidence. Volcanic activity, the formation of mountains, and the action of glaciers and rivers, tides and ocean currents also became better understood. This series includes landmark publications by pioneers of the modern earth sciences, who advanced the scientific understanding of our planet and the processes by which it is constantly re-shaped.

Memoir of Leonard Horner, F.R.S., F.G.S.

Leonard Horner (1785–1864) was a prominent geologist, educator and, later, a factory inspector. In 1833 he was appointed to the Royal Commission on the employment of children in factories, and he inspected sites around the north of England. His earlier scientific work saw him elected a fellow of the Royal Society in 1813, and he was twice president of the Geological Society. The two-volume *Memoir of Leonard Horner*, edited by his daughter, Katharine Lyell, and published in 1890, is a selection of letters to and from his family and friends. The correspondence gives vivid insights into the world of this influential reformer. Volume 2 covers the period 1839–1864, and includes letters about Horner's activities at the Geological Society, his travels in Italy in 1861, his political concerns, and key moments in his personal life, such as the birth of his grandson in 1850.

Cambridge University Press has long been a pioneer in the reissuing of out-of-print titles from its own backlist, producing digital reprints of books that are still sought after by scholars and students but could not be reprinted economically using traditional technology. The Cambridge Library Collection extends this activity to a wider range of books which are still of importance to researchers and professionals, either for the source material they contain, or as landmarks in the history of their academic discipline.

Drawing from the world-renowned collections in the Cambridge University Library, and guided by the advice of experts in each subject area, Cambridge University Press is using state-of-the-art scanning machines in its own Printing House to capture the content of each book selected for inclusion. The files are processed to give a consistently clear, crisp image, and the books finished to the high quality standard for which the Press is recognised around the world. The latest print-on-demand technology ensures that the books will remain available indefinitely, and that orders for single or multiple copies can quickly be supplied.

The Cambridge Library Collection will bring back to life books of enduring scholarly value (including out-of-copyright works originally issued by other publishers) across a wide range of disciplines in the humanities and social sciences and in science and technology.

Memoir of
Leonard Horner,
F.R.S., F.G.S.

*Consisting of Letters
to his Family and from Some of his Friends*

VOLUME 2

EDITED BY KATHARINE M. LYELL

CAMBRIDGE
UNIVERSITY PRESS

CAMBRIDGE UNIVERSITY PRESS

Cambridge, New York, Melbourne, Madrid, Cape Town, Singapore,
São Paolo, Delhi, Dubai, Tokyo, Mexico City

Published in the United States of America by Cambridge University Press, New York

www.cambridge.org
Information on this title: www.cambridge.org/9781108072854

© in this compilation Cambridge University Press 2011

This edition first published 1890
This digitally printed version 2011

ISBN 978-1-108-07285-4 Paperback

Affectionately Yours
Leonard Horner

MEMOIR OF

LEONARD HORNER,

F.R.S., F.G.S.

CONSISTING OF LETTERS TO HIS FAMILY AND FROM SOME OF HIS FRIENDS.

EDITED BY HIS DAUGHTER

KATHARINE M. LYELL.

IN TWO VOLUMES.—VOL. II.

[PRIVATELY PRINTED.]

London :
WOMEN'S PRINTING SOCIETY, LIMITED,
GREAT COLLEGE STREET, WESTMINSTER, S.W.
1890.

CONTENTS OF VOL. II.

CHAPTER I.
1839—1840.

CHAPTER II.
1841.

CHAPTER III.
1842—1843.

CHAPTER IV.
1844—1845.

CHAPTER V.
1846.

MEMOIRS

OF

LEONARD HORNER.

CHAPTER I.

1839—1840.

To his Wife.

Manchester, *4th April,* 1839.

I WENT on Tuesday to Mr. Samuel Greg's. I met with a
very kind reception from him and Mrs. Greg, whom I had not
seen before, and from all his people. She is a gentle, sensible,
pleasing person. I yesterday went to the factory with him,
saw some of the people, and then visited a capital new school
he had built since I was there, and some beautifully neat and
comfortable cottages, models of their kind. When we were at
tea between eight and nine, I heard the sound of music under
the drawing-room window, and Mr. Greg told me that the
worthy man Bancroft (one of his workers, whom you will.
recollect me as describing as teacher of the Sunday School,
and preacher), has got up, since I was there, an excellent
singing class for sacred music, and that out of respect, and to
pay a compliment to me, he had brought up the class to sing
under the window. On looking out, there I saw Bancroft

A

with his broad brim hat, holding a music book, a man playing a violoncello, another a serpent, boys holding lights, and about thirty young women standing in a semi-circle singing a Psalm very beautifully. I thanked them most cordially for their kindness. It had been done, of course, with Mr. Greg's knowledge, and he had desired a fire to be made, and the lamps lighted in the Sunday school room near his house, and tea to be got ready for the party. As it was a bitter cold night, I begged that they would go immediately to the school-room, and I should be glad to meet them there, and hear some more music. Mr. Greg and I accordingly went, and found them assembled. I went round among them, and said a few words to some, whom I recollected to have seen two years ago. They had tea and currant loaf, and then sang two anthems and a chorus exceedingly well. They were very nicely dressed, and the Members of the Silver Cross (the order of merit instituted by Mr. Greg) wore their insignia. I said a few words thanking them for the honour they had done me, which I assured them I felt very sensibly. It was very kind as a spontaneous act on their part and no less kind on the part of Mr. Greg to approve of it. I left them this morning after breakfast, and was much gratified by my visit, and I came away confirmed in my opinion of the sense, benevolence, and enlarged philanthropy of my friend. We may be on a barrel of gunpowder here, but there is no out-ward sign to cause alarm. Mark Phillips says he is not at all easy about these secret warnings, and thinks our prospects generally bad.

To his Daughter.

Manchester, *9th April,* 1839.

MY DEAREST KATE,—I went yesterday to Claremont, nobody was there but Sir Benjamin and Lady Heywood, and their aunt, Miss Percival. I spent a quiet agreeable afternoon and left them this morning after breakfast, and had a long walk round by Swinton and Eccles, visiting mills. I passed Patricroft, the great foundry establishment of Mr. Nasmyth of Edinburgh. His father taught your aunts to paint in oils, and he is still alive, above eighty, and painting still, his son tells me. I called on Miss Nasmyth, and found her painting a very pretty landscape. He is an excellent young man, very industrious and prosperous in the world. He was an early pupil of the School of Arts, and is now making locomotive engines for railroads. Most affectionate love to all.

God bless you, dearest Kate,

Your affectionate father,

LEONARD HORNER.

To his Wife.

Newby Bridge, Windermere, 5*th June*, 1839.

I got through my business at Lancaster satisfactorily, having large, clean, airy places to visit, nice, tidy, comfortable looking people, and reasonable, well-disposed masters. In the evening I walked to the Castle and surveyed the distant mountains over Morecombe Bay, and the rich verdant fields that surround the Castle. Ingleborough was enveloped in mists. This morning was most propitious, so I planted myself on the top of the Ulverstone mail, and had a most delightful drive to this place. We started at mid-day, a bright sun, with a cloudless sky, and a most delightful air. Near Haversham is Levens Hall, a very ancient family mansion, at present in the possession of Colonel Howard. It is surrounded by a garden of great extent, with yews cut in all sorts of fantastic shapes, and stands in a park with many noble trees. From hence is an extensive moss in the low-land watered by the Trent. They were busy casting their peats, and glorious weather they had for it to-day, and indeed the dryness for some time has been very favourable for that operation. A considerable part of the moss has been pared off, and the sub-soils afford the finest corn land, for there is below a deep bed of marl, which they bring up, and cast on the surface. It is curious to see the finest corn fields in the midst of a peat bog. The road winds along the foot of Whitbarrow-Scar, very lofty cliffs of mountain limestone, and the village of Lindal, where we enter upon the slate rocks, and the character of the country quite changes as we enter then into highland scenery.

To the right of the road from Lancaster, and near Burton, there is a limestone hill called Farlton Knot, in which the strata, quite bare and exposed, are seen to dip to the east, while Whitbarrow presents the same appearance, only dipping west, so that if prolonged, the rocks would form an immense arch, the two escarpments are at least six miles asunder, but I have no doubt they were once continuous, but have been broken down; in the plain between, small ridges of the strata set on edge, stand up every now and then, like islands in a lake; in some places there are long continuous and wooded ridges, which put me strongly in mind of the great stream of lava from Puy la Vache, which we crossed in going to poor Count Montlosier's house.

I got here between three and four, and have found a most capital Inn, everything delicately clean, and civil people. I noticed a view of mountains in the room and the landlord said, "Yes, sir, those are the Alps as seen from the Jura, I brought the plate from Geneva." "Vous parlez Français peut-être? O oui Monsieur—e anche un poco Italiano." He had travelled long with his last master, and as Martha, his wife, had also been in a good family, they know what gentlefolks like, and are accustomed to.

They gave me trout just out of the lake for dinner, and I soon after sallied forth for a walk. Newby Bridge is about a mile from the lake proper; the Leven which flows out of the

lake is running past the window at which I now write. I got
to the margin of the lake by a road flanked with wooded cliffs
on one side, the bright fresh hazel leaves contrasting with
many dark sombre pines. It was a lovely evening, warm,
bright, not a breath of wind, the wide lake spread out like a
mirror, the birds singing most joyfully. While standing in
this peaceful scene, I thought of Charles then dining at the
Geological Club in the Strand, and how much better off I was,
although I should have been very happy had I been there.
But I felt a sad want (we are never contented), *I was alone.*
I longed for you and my dear girls to be at my side, that I
might myself say, and hear you and them say, " how beautiful," " how peaceful! " I recovered myself from my
unreasonable fit, and said, " Well I will go back and pour
forth all I should have said, over a sheet of paper." I am going
to take some tea, and have another walk before the sun has
quite gone down. . . I have had my walk of a couple of miles
and I found out a very pretty winding by-road. A man of the
inn was fishing, and had just caught a most beautiful char of
unusual size. They are very scarce, and it is an invariable
custom here, when char is given for dinner, to charge a shilling
apiece for each char, great or small. It is half past nine and
I have only just got candles, so much for being so far to the
north. It is a fine settled evening, and they tell me it bids
fair for good weather to-morrow, and I am looking forward to
a delightful row up the lake to Lowood.

I have been reading since I came here the " Eclogues " of
Virgil, it is a very long time since I read a Latin author,
and I am delighted to find that I have lost very little, and
still more that it gives me so much pleasure. It is true that
I have a Delphin edition, which is a considerable help. It is
like recovering an old friend, and I will not lose him again if
I can help it. I read this book while I was at breakfast and
for half an hour afterwards, while breakfast was digesting. I

hope to get through the four books of the "Georgics" before I
return home, I have been refreshing my memory and getting
many new ideas about the geology of the lake district, by
reading these memoirs of Sedgwick which I brought with me,
and if the weather is fine I mean to give myself a day of
geology, for there are some capital facts described by Sedgwick
within an easy distance of Lowood.

To his Wife.

Low-wood Inn, *June 6th*, 1839.

I rose before six, it was a beautiful morning; I walked a
mile and a half to Back-barrow, a beautiful spot on the Leven,
where there are large cotton works belonging to Mr. Ainsworth,
a great friend of Mr. Wood, the chairman of the Board of
Excise. They are in high order, and there is a school for the
children, seventy in number, which is doing well.

I believe that this good I have accomplished, for I found
the schooling part very bad last year, but having in truth a
good man to deal with, my recommendations have been
attended to. He asked me to breakfast with him ; he is a
bachelor, but I found a sister living with him, a very lady-like
person.

At ten o'clock I started from Newby-bridge in an excellent boat,
and had a delightful row, the whole length of Windermere to
this place. After dinner I started with a guide and my
hammers to explore some things described by Sedgwick. I
had a glorious walk of six miles, ascending some hills about
eight hundred feet above the lake, and I think I had the
finest view before me that I have yet seen in any part of the
lake district. Windermere spread out in its whole extent with
all its islands and gorgeous banks, with some graceful sailing
boats on its azure surface ; the vale in which Brothy is
situated with the silver stream of the Rother meandering

through emerald meadows, and the sweet retired grassy vale in which Ambleside lies, and the grand mountains of Langdale, Pikes, Bowfell, and others without names, bounding the view, a superb sky to set all off to the best advantage.

To-morrow I go to Keswick, but we shall have rain, for heavy clouds rest to-night on the mountains.

To his Daughter, Mrs. Lyell.

Keswick,, *June 9th*, 1839.

MY DEAREST MARY,—I have made the acquaintance of Charles' friend, Mr. Thomas Spedding, and I dined there yesterday, and what I have seen of him I like much; he is so gentle and benevolent in his opinions. The Magistrates' clerk, Mr. Bates, who showed me the way to his house, said that Mr. Spedding is much beloved here, and does much service to the poor as a magistrate, by giving them his advice, and keeping them out of litigations. There were Archdeacon and Mrs. Headlam, also agreeable people. He has asked me to visit them when I am in the neighbourhôod of Barnard Castle, where he lives. Mr. Spedding's house, Greta Bank, is beautifully situated about a mile from Keswick, at the base of Skiddaw. He said that Charles had promised that you would pay them a visit; he said that if he could get Charles entangled among the rocks here, he would have a good chance of keeping him. I told him that they lie far too low down in the geological series to make an early visit probable. They have been kind enough to ask me to dine with them again to-day, which I mean to do.

My kindest love to Charles and all,

Your affectionate father,

LEONARD HORNER.

To his Wife.

Cockermouth, *20th June*, 1839.

Our wedding day, the thirty-third anniversary. I was sad

that we should be separated. I hope that better luck awaits us, and that we shall pass many more of them together.

I think you and the girls would find it very instructive and amusing to go more frequently to the British Museum. The advanced civilization of the Egyptians at a very remote age, is very remarkable. I believe the date of the Pyramids is not less than three thousand years before the birth of Christ, and what ages must have elapsed from the time when the country was first inhabited by savages, until they could have arrived at that degree of perfection and refinement which would enable them to erect such wonderful structures. There is a well-grounded conjecture, I believe, that they received their civilization from India. If you can borrow the recent work of Wilkinson on the domestic manners of the ancient Egyptians, you would find it very interesting.

[In July Mr. Horner was much engaged in business connected with the Factory Bill which was going through Parliament. In August he went to the meeting of the British Association held at Birmingham, and met the Charles Lyell's, who had arrived from Scotland.]

To his Wife.

Birmingham, *August 27th,* 1839.

Yesterday evening we went to the Town Hall, where there was a general meeting of the Association, when the President, the Rev. W. Harcourt, son of the Archbishop of York, read a very excellent address. He dwelt at considerable length upon those attacks that have lately been made upon men of science, and especially geologists, as impugners of the Christian faith, and he confuted these admirably well. It is most creditable to him as a clergyman to have made the stand he did. To-day there is a conversazione in the Town Hall. Sir Robert

Peel has asked two parties, one for to-day, and another for to-morrow, at his house, Drayton Manor, near Tamworth, to dine and stay all night. I am included in the party for to-morrow.

[Mr. Horner returned to his district in September, and he determined to have two of his daughters with him to lessen the trial of separation from his family, Mrs. Horner feeling it right to remain at home with her other daughters.]

To his Wife.

Manchester, *October 4th*, 1839.

My two dear companions give you in minute detail, an account of all our proceedings, so that no topics are left for me. They are dear good girls, delightful companions, and contribute very greatly indeed to my comfort; nothing can exceed their affectionate attention to me. I trust they will not weary, but they have such ample resources that I am not afraid of that. There are very dismal forebodings in all the cotton mills that they will only work two-thirds time, which will of course cause a great reduction in the wages of the people.

5th October. I have begun to read to the girls Buckland's "Bridgewater Treatise," and have read about two hours. It has been a very happy evening to me to be thus sitting with those two dear girls.

Next Wednesday is the birthday of our dearest first-born, and I have been thinking of our present to her.

Sunday morning. A fine day, and I am going out with the girls to the Collegiate Church. God bless you all.

To his Daughter.

Manchester, *20th October*, 1839.

MY DEAREST KATHERINE,—I would have written to you before now had it not been that my dear companions take away all my subjects in relating what passes in common, and my own occupations afford nothing but materials for official despatches.

Frances and Susan are at this moment sitting at the same table with me, and their pens, particularly that of Susan's, move with a velocity which I cannot pretend to, they have much to say that will amuse you all. I had great pleasure in taking them on Thursday to Barlow Hall. The dinner was very agreeable to me, for I was well placed between Mrs. Phillips, whom I always like to sit next, and a charming Mrs. Tomkinson, one of the most sensible and most graceful persons I have met with for a long time ; I was glad to renew my acquaintance with Colonel Fergusson, Sir Ronald's son, whom I had not seen for several years. Mr. and Mrs. Phillips kindly asked me to come again next day, to go to the ball, but although it would have been gratifying to me to see your sisters there, I did not feel inclined to go and return alone to Manchester very late at night. They appear to have enjoyed everything very much, and to have met with great kindness. I went to Rochdale on Friday morning, and returned to dinner, but to a very different room, and I felt very uncomfortable in my solitude. I am spoiled for living alone, and I do not know how I shall reconcile myself to it again. I consoled myself with " Nicholas Nickleby." I began it about a week ago, taking a number in my pocket, which I read as I am travelling, or when I have a vacant half hour while the factory people are dining. It is in many parts clever and amusing, but how very inferior he is to Scott. There is scarcely a character or an incident that is not a caricature and extravagant, whereas in Scott everything in all his characters and events, is consistent with nature and probability. Dickens is obviously unacquainted from personal experience with the higher ranks of society, in depicting them he is evidently not at home, whereas from the tradesman downwards, he has it is clear had the most ample means of observation ; and has made good use of his opportunities. I was very happy to see their faces next day, and I took them to visit Mrs. Wood at Singleton Lodge, about three miles on the opposite side of Manchester from where we live, as soon as they came back, as it was a fine day, and fresh air and exercise was good for them after their dissipation.

I am, my dearest Kate,

Your affectionate father,

LEONARD HORNER.

To his Wife.

Manchester, *Nooember* 2nd, 1839.

I am glad to hear so good an account of Whishaw ; as soon as I return I shall go to see him. I had a long and interesting letter from Mallet yesterday, for which pray thank him the first opportunity you have. Ask him who it was made the speech to Hallam about his late book which he quotes : " Why, my dear Hallam, what a quantity of trash you must have read." We have puzzled over the name in Mallet's most un-official hand, but cannot make it out.

I hope you went to Norwood, if you did I am sure you would be interested by the school, especially if you were fortunate enough to meet Dr. Kay there.

The weather here for some days has been excessively cold, and as there has been much wind, it was most piercing. The dust has been very unpleasant. It was curious to see at this time of the year in Manchester, a cart engaged in watering the streets. Happily the month is arrived before the end of which I hope we shall be again united. Although I have not the less counted the days gone, and how many remain, I have felt totally different this visit from what I ever did before, by having the society of those dear girls, and the home feelings

which their presence creates.   We get on most merrily
together, and we have frequent talks you may believe about
dearest Mamma and all those that surround her.

To his Wife.

Manchester *April* 21st, 1840.

I had a prosperous journey and I read nearly the whole of
the first volume of Sir Samuel Romilly's Memoirs.   I found
it extremely interesting, particularly all his correspondence
and observations about the early part of the French Revolution.
Madame Delessert appears to have been a remarkable person,
but her daughter, Madame Gautier, still more so, some of her
letters are admirable.   I have been disappointed by those of
Mirabeau.   It is charming to see the lofty, pure sentiments
that breathe through all Romilly's letters.

To his Wife.

Manchester, *April* 30th, 1840.

You may believe how greatly I have been afflicted by the so
sudden and unexpected death of our amiable and excellent
friend Mrs. Hallam ; your letter was not the first announcement of it, for ha an-hour before I had read the notice of
it in the newspaper.   Poor Hallam ! what a succession of
afflictions he has been visited with in a few years ; he seems
only to regain his cheerfulness to be struck down by another
blow ; and poor Julia, and her poor father, Sir Abraham Elton,
now ninety years old.   She was so cheerful, so motherly, so
sensible, so unobtrusive, so humble; I shall never have another
of those pleasant breakfasts ; we had planned to go to see
some sights together as soon as I came back ; I shall write to
Dr. Holland to ask about Hàllam, and to convey to him my
most heart-felt sympathy in his affliction.   I applied to Lord

Palmerston to get for me the Prussian law for the regulation of the labour of children, and Mr. Backhouse, the Under-secretary, sent the papers he had received from Berlin, to me to-day. They are written in the German character, which I am so little familiar with, that it would greatly waste my time to decipher it, so I send the three papers, and request that some one of my five learned clerks,* will undertake the task of making a copy of each in our ordinary written character. I should like this to be done as early as convenient.

You remember my interview with the Bishop of London, and my application to the National Society to relax their rules in favour of factory children attending their schools, by not requiring them to learn the Church Catechism and Church formularies should their parents object on religious grounds. I have had a letter to-day from Lord Ashley, telling me that yesterday "the resolution had been carried *unanimously* in a very full meeting of the Committee." This is very gratifying to me, as I cannot but consider it a step gained towards a better feeling on this vital question ; Lord John told Dr. Kay, that he feared I should not succeed. To-morrow I hope to have an uninterrupted day of work at my pamphlet, except going out on one business. I am making an inquiry with the view of finding out whether it be possible to frame a clause for the new Factory Act that can be enforced, to prevent or diminish the risk of those frightful accidents from the machinery which are so constantly happening, by not enclosing the dangerous parts. It is a very difficult matter to deal with, but so far as I have gone in my inquiry my hopes of being able to do something are great. Frances asks when the Factory Bill was first proposed in England. As far back as 1802, and there were eight bills between that and 1833, when the present law was passed.

* His daughters.

To his Wife.

Rochdale, 28*th June*, 1840.

I was yesterday engaged with my usual duties : dull and mechanical and unvaried.   I recover my spirits by thinking that I am the instrument of making the lives of many innocent children less burthensome, and that by earnestness in my calling I may acquire that degree of credit, that my object of extending the protection to children more widely, will be more attended to.   I was at a factory yesterday where there are fifty children employed, and I think that I made some impression on the masters in favour of my plan of getting the children's work reduced to half a day.   Happily it falls in with the interest of the master, where a sufficient number can be had to make out the morning and the afternoon set.   Were this general, and fixed by law, and if good schools were spread over the land, what a change would be seen in the next generation !   I think there is a better feeling manifesting itself among these masters here, who have hitherto been very unfriendly to the Act.   I am to be another day here, and on Tuesday I strike my tent to put it up again for the rest of the week at Oldham.   This is about the most barbarous part of Lancashire, at least they have the reputation in other parts of the county of being very rude ;   and certainly there is not much external grace among them ;   but whose fault is it that they are so ?   not their's, but that of the better classes around them who have neglected them.

I have been reading the " Lebens Nachrichten ueber Niebuhr," and am much interested by it.   The volume I have is that which begins with his leaving Rome in 1823, and then describes his settling at Bonn, and the rest of his life, which ended there in January, 1831.   The letters are so attractive that I can hardly lay down the book.   It is very pleasant to have the old places and our old friends recalled to mind.   He speaks of Brandis with great affection, and as certainly the

most distinguished man of the University. He got to Bonn
in September, 1823, and in none of his letters up to January
1825, does he mention either Schlegel or Windischmann. The
letters hitherto have been only to his sister in Holstein, whom
he seems to have had the greatest love and respect for, and to
his wife, when he was absent from her for some months at
Berlin. The warmth of his affection for her and for his
children is quite beautifully displayed in the minuteness and
earnestness with which he enters into all their occupations,
and all their pleasures and pains. His eldest boy, Marcus,
I am quite interested in, and have become attached to; if he
is alive he must have the greatest pleasure in reading these
letters over and over again. But he, Niebuhr, appears to have
been of a most sensitive nature. In politics he has a horror
of liberalisms, and goes far beyond what one might have
expected of a man of his understanding. It is very early for
me to pronounce an opinion upon him, but he seems to be
more adapted to shine as a man of letters, than as an active
politician. He speaks in such terms of the Crown Prince, the
present king, that he certainly must possess many excellent
qualities, both of the head and heart.

You remember, I daresay, that when we were at Bonn,
during the Crown Prince's short visit there, he went with
Professor Holweg one morning before breakfast alone to visit
poor Niebuhr's grave. I read German, and every thing about
Germany, with a pleasure that I never feel in reading French,
and nothing I should like better, than to pass some time again
in *lieben Deutschland.*

To his Daughter.

London, *July 5th,* 1840.

My Dearest Susan,—This morning I was at work at half
past six, to bring up some arrear of business, and at eleven I

went to call on Lord Ashley by appointment, first to talk about
the poor girl who has been so sadly mutilated by the machinery
of a mill.   He is a truly benevolent man, and it is by no
means talk alone, as I have proof in this case.   He has put
me in communication with his own solicitor, and he told me
that he will have the best counsel on the circuit employed to
obtain redress for the poor girl, and to expose the frightful
dangers to which these poor people are exposed by the care-
lessness of their employers.   My next business was to talk
about the motion of which he has given notice, of an address
to the Queen, for a Commission to inquire into the employment of children in other works.   This is quite in accordance
with the views I have expressed in my pamphlet.   I have had
a conference with Mr. Chadwick and Dr. Kay, who are quite
willing to give Lord Ashley all the help they can, and to bring
the immense staff of the Poor Law Commissioners, about 4,500
persons, over the kingdom, to aid in the inquiry.   Having
talked with Lord Ashley about it, I have arranged for a
meeting to-morrow with his Lordship, Chadwick, Kay and
myself, to consult about the best way of preparing for the
Commissioners.   It is a happy thing to bring men of different
politics together, to unite in such a work of humanity and
justice, and sound policy.   I have seen Mr. Lefevre too, the
Poor Law Commissioner, and brother of the Speaker, and he
is quite favourable to the enquiry, which is very important,
because he will speak to Lord John about it.

I am, my dearest Susan's

Affectionate father,

LEONARD HORNER.

To his Daughter.

London, 14*th July*, 1840.

MY DEAREST KATHARINE,—On Friday I went to Hampstead
to dine with the Mallets, no one was there, and it was there-

fore more agreeable, and they asked me to return next day, and stay till Monday, which I did. The garden is in the greatest beauty, as well as those on each side, and on Sunday the air was so clear, that the round tower of Windsor Castle was quite plainly seen, and I saw distinctly the Devil's Punch Bowl, beyond Godalming on the Portsmouth Road; I had a walk by Caenwood lane, and through the fields, which I enjoyed very much. I had "Hermann and Dorothea" in my pocket, and read the greater part of it. I had not read it for three or four years, and like it not only better, but am surprised and ashamed that I did not like it very much before. I see by the advertisement that Wilhelm von Humboldt wrote "Eine Æsthetische Versuche" upon the poem, which I should like very much to read.

The proceedings against the mill owner, at whose mill the poor girl, Mary Howorth, was so sadly mutilated (as I mentioned in a former letter), are begun; and the action is to be brought by Lord Ashley as her friend, in order that the father may be brought forward as a witness. He told me that he has retained Cresswell, the most eminent Counsel of the Northern Circuit. Come what will of it something will be done for the poor girl, I hope the fellow will be obliged to settle an annuity upon her. I am happy to say that there is a prospect of a reconciliation between the heads of the Church and the Committee of Privy Council on the education question, by mutual concessions. Lord Lansdowne told me this on Saturday, when I was calling upon him to talk about Lord Ashley's motion. He said he was to see the Archbishop of Canterbury that day, and Dr. Kay told me that the meeting went off very satisfactorily, and that there is a reasonable prospect that some great and united effort will be made in the cause of national education. This is very cheering.

To his Daughter.

London, *2nd July*, 1840.

MY DEAREST LEONORA,—Yesterday I went to the Athenæum and dined early, Lord Nugent joining with me in partaking of a joint of cold beef, and he talked to me a long time about the abolition of the punishment of death, on which subject he is writing a pamphlet. I then went down to the House of Commons to hear Lord Ashley bring forward his motion for an enquiry into the employment of children in various trades. I saw him in the lobby, and he told me he had some very strong cases to bring forward, and read me a most melancholy account of about 500 children employed in pin making at Warrington. Most unfortunately, when the Speaker came, there were only 36 members present, so the House was immediately adjourned, for the Speaker cannot take the Chair unless 40 be present. Not five minutes afterwards there were at least 50. Thus Lord Ashley lost his opportunity, and it is difficult to say whether he will find another open day this Session. I saw him afterwards and he was very much annoyed, and blamed the Speaker very much. The House had met in the morning, and it was not supposed that, when it resumed at five, it was to be counted again, otherwise he would easily have had the requisite number; the Speaker told him the evening before it would not be counted, but he changed his mind afterwards, and did not give Lord Ashley notice. Several members spoke to me about it. As I was talking with Lord Ashley, Mr. Maule came up to him and said that the Government did not mean to oppose his motion, which renders the loss of the opportunity still more provoking. I have had a note from him this afternoon, in which I see he is very chagrined. I hope, however, he will still find some open day. I came home about seven, and passed a quiet evening, very much interested in Mitchell's "Life of Wallenstein."

Your affectionate father,

LEONARD HORNER.

To his Wife.

Leamington, *9th September*, 1840.

I arrived at Coventry at half past one and got into the
Warwick mail, and was soon in Galton's house. They are all
very well, Mrs. Galton better, and looking better than she has
done for some years. The last account they had from Francis*
was from Pesth, and he was to set out next day for Constanti-
nople. They read me his letter, which is an admirable one,
evincing great powers of observation. They also read me
another letter from him giving an account of an accident
which very nearly cost him his life. You remember at the
Oxford and Cambridge rowing match, your seeing the collision
of a steamer, and three boats run down, he was on board the
steamer, and by the shock was thrown headlong overboard,
under the paddle, and had he not been a very skilful swimmer
and diver, and possessed of rare courage and presence of mind,
he must have perished, from the description he gives of the
situation in which he was. He does so, not at all boastingly,
but playfully. We have seen him more than once since, and
he never said a word about it, that I remember. They are a
little uneasy at his travelling in those countries in such hot
weather, all alone.

Tertius and I walked to the pump room before breakfast
this morning.

———

[Mr. Horner was appointed, with Mr. Saunders and two other
gentlemen, on a commission to enquire into the employment
of children whose labour was not under any restrictions, on
the 20th October, 1840.

The first report of this commission was dated the 16th
November, 1841, and the result of the enquiries and recommendations of the commissioners, was the passing in 1842 of
the " Mines Inspections Act," whereby the labour of females

* Author of " Hereditary Genius," &c.

in mines was prohibited, and restrictions were placed upon
the employment of boys in mines, and in 1845, of the Print
Works Act, whereby these establishments were placed
under inspection.]

From Mr. Hallam.

Henley Park, *16th November* 1840.

MY DEAR HORNER,—I found your letter on Saturday, when
I returned from my short visit to the north. As soon as I
found that you were not to be at Manchester, I gave up all
idea of taking that town into my present scheme, deferring it
to some future opportunity when I may have the advantage
of your guidance. I passed three days with the new dean at
Spofforth, his living in Yorkshire, where I met Longley, the
Bishop of Ripon, and his family. He is a very agreeable man,
and justly popular in his diocese. The Murchisons had been
invited, but are detained in Scotland. I spoke of you to
Herbert, and have made you acquainted with him by
anticipation. Do not fail to call; he is now at Manchester,
in a house belonging to the late warden, which he has taken
for a short time: he finds houses very difficult to procure in
any tolerable situation, or convenient in themselves, and
complains that the manufacturers build large reception rooms
with little other accommodation. He has been most
satisfactorily received by all parties, and is, I am confident,
desirous to exert himself, and do all the good in his power,
not inclining to any party, though he is strong in his political
opinions, and his wife more so. His daughter is sensible and
unaffected—if you have any of yours with you, they would
probably like each other. Herbert has very little experience
of mankind, and least of his own profession, knows nothing
of the squabbles which distract the large towns, and must
therefore walk cautiously, which he is inclined to do. On the
whole I found him much more disposed to buckle to his work

than I had expected, he has preached often, and ιo *crowded houses.*

I went from Spofforth to Trentham, where a friend of mine has the living, the weather generally bad. I came in the worst day of the year, last Friday, to town. I had designed to go the preceding day, in which case 1 should have come in for the disaster near Harrow, though it would not have been more than a temporary delay to me. I stayed by persuasion over Thursday to dine with the Duke and Duchess of Sutherland, whom I know very slightly. They have a daughter, Lady Elizabeth Gower, of remarkable character for intellectual attainments, and also for goodness.* She is but seventeen. Her parents are very worthy people, and with no appearance of worldliness.

I shall now continue here till over Christmas. It would, I fear, be vain to ask you and Mrs. Horner, but if you have a day to spare, my sister and I shall be most glad to receive you. One hour to Tringford, and my carriage shall meet you there. But I am forgetting that you may be gone already to the north.

Yours very truly,

H. HALLAM.

---

* Married to the Duke of Argyll in 1844, died in 1878

CHAPTER II.

1841.

To his Wife.

Stalybridge, 18*th May*, 1841.

As I shall be out all day to-morrow at Mossley, a village on the confines of Saddleworth, I write these few lines to-night, to tell you that I am passing my time in the uniform routine. Smyth's " French Revolution " is my *delassement*, and I find it very interesting and instructive. I am, however, going a little beyond my strictly official duties, and am trying to do something to promote education in this most strangely neglected place, where gain appears to have been, and to be, the only thing thought of. The population of Ashton, Stalybridge, and Dukinfield, (three places which are contiguous as to constitute one vast town), amounts, it is believed, to sixty thousand; and there is not at this moment a single endowed day-school, neither National or Lancastrian. A National school is now building at one extremity, " but what is that among so many "?   In Dukinfield, where there are about twenty thousand, there has not been up to this hour a single place of worship open, belonging to the church establishment; a new church that will hold about one thousand five hundred, has been built, and is to be consecrated next Monday.

I have written to the committee of the National Society, to call their earnest attention to the destitution of this place, and I am going to write to the Committee of the British and Foreign School Society, and to the Committee of the Privy Council.

I am very happy to hear of the intention of the Powers* to visit their native country ; and am glad that you have written to say how truly glad we shall be to see them. I shall write to ask them to postpone their visit till towards the end of June, that I may see them.

I am glad to hear that the Horticultural fête turned out so successfully.

——

[In June, Mr. Horner went to pay a visit to his old friend Lord Murray in Edinburgh.]

——

To his Wife.

Edinburgh, 9*th June*, 1841.

When I closed my letter to Joanna yesterday, Murray and I went out at half-past five in the open carriage to Craig-Crook. I never saw the place in such beauty, the trees have grown much all around it, and the foliage and grass is in the most brilliant state of verdure ; several thorns covered with blossom interspersed. We found Jeffrey better ; he gave me an account of his sudden attack in the court, and it was evidently an affection of the stomach, and nothing more. He was revived from his fainting, or rather faintness, by a table-spoonful of brandy and water. He came to town to-day with Mrs. Jeffrey, and I saw him looking better than yesterday.

We were quite alone at dinner, but Lady Murray and I went to drink tea with the Bells.† They are both looking well, but she in particular, really as young as ever. George, his brother, was there, and I am sorry to say that he is very much aged since I saw him ; he is not totally blind, because he said to me, looking upon the view from the window, " I can still see that, but I can neither read nor write." He was

———

*   His sister, and her husband.
†   Sir Charles and Lady Bell,

placid as usual, and cheerful, and appears to be bearing his great misfortune with great philosophy.

Murray and I sat talking till twelve, and when I went to my room I saw the sun travelling along the horizon towards the East, by the bright twilight on the horizon. I slept well, but I was up again at six, and the view was finer than yesterday, for there was less mist. The coast of Fife was clear, and the Forth was a bright blue, and the vessels with their white sails were scattered upon it for a long way. This is a capital house, beautifully furnished and fitted up. I occupy Mr. Murray's * bedroom, which, by choice, he made at the top of the house, and you can conceive how fine a situation it is. The new Episcopalian chapel close by the draw-bridge, is a new and good feature in the landscape, I am now sitting in the room with the window open, and a bright sun shining upon the green fields.

We breakfasted at half-past eight, and Murray soon after went to Court, and I set out shortly afterwards. I called on Macbean, and then across to Cockburn, who was gone to the Court. I was desirous of paying one of my first visits to Pillans, with whom the world has gone very hard lately, as you well know. I found him looking much better than I expected, indeed well. Coming up from London Street, an old friend recognised me, as I recognized him, a chairman standing at the corner of York place, Hector Maclean. "You have been a long time standing at this corner," I said, "I daresay more than thirty years." "Aye," he replied, "forty-three, and you'll mind I cleaned your gun when you were a volunteer ! " I gave Hecter half-a-crown.

I next proceeded to the parliament house, and went first to the room where Cockburn was sitting, with counsel pleading before him : Maitland was one of the counsel, but as he was not speaking, he came out to me, and we had a little talk.

* William Murray, Lord Murray's elder brother.

He was in his silk gown as Solicitor General, and looked well. Cunningham was in another Court, and I peeped at Murray on the judgment seat, in another place. I spoke to some other friends in the outer-house, as you know the great hall is called, and then went forth to view the old town. I called on the worthy Bryson to speak about the School of Arts, then along Princes Street, looked into the Royal Society's Rooms, and then called on the Thomsons.

I must break off suddenly, Murray has asked me to walk out. We have been to the beautiful gardens along the water of Leith behind the house, and I have only ten minutes to dress for dinner.

God bless you all.

To his Wife.

Edinburgh, 10*th June*, 1841.

I bless for the thirty-fifth time the return of this day, to me the most blessed day of my whole life;* and I have to add for the eighteenth time the wish, that my dear Joanna may see many happy returns of this same tenth of June. I trust that you are both at this moment as happy as I wish you.

I broke off my letter yesterday suddenly, and now resume my story at a time less liable to interruption; it is seven o'clock in the morning, and I am sitting with my window open looking over the beautiful country; it is a bright sunshine, scarcely a breath of wind, and what there is, is from the West, so we shall probably have a warm day. Well, I went to Cockburn's at two, and we bent our steps to the Academy. We got there before the school broke up for the day, and sat half-an-hour in one of the classes, and I never saw fifty finer or more gentlemanlike boys collected anywhere. It was the class of Mr. Macdougall, a new master since we left Edinburgh, and they tell me that after the Rector, he is the best. He appears a clever man, and from the little I saw, he

* His wedding day, and youngest daughter's birthday.

has an excellent method. I saw the Rector, and some of the other masters. Williams says they have three hundred and fifty pupils, which he says is as many as they can do full justice to. Cockburn says that everything is going on admirably well, in the school, with the masters and with the directors. I looked upon the flourishing state of my work with great satisfaction, and there has been no alteration in the system we established seventeen years ago. We next strolled about that neighbourhood, and finished with the new railway to Glasgow, which will be open this autumn. We dined without any addition to our party, but in the evening, they asked the Cockburns, Maitlands, Pillans and Macbean to come, and they went away at twelve ; Cockburn exactly as he used to be, merry and joking.

To his Daughter.

Edinburgh, 11*th June*, 1841.

MY DEAREST SUSAN,—I had an appointment yesterday with Lord Cockburn at two, and we went to Morlands to see Dr. Thomson. He is an invalid in body, but is in full vigour of mind, and it was very pleasant to see his energy, and the lively interest he takes in all that is going on. Dr. Allan Thompson, now a Professor at Aberdeen, is on a visit. I passed by Whitehouse, and thought of past days. Old Pitfodels,* the founder and supporter of the Nunnery, was going in at the door, accompanied by a young priest. We looked in at Sir George Warrender's, and the avenue was in great beauty. We had at dinner yesterday, Mr. Primrose, Lord Rosebery's brother, and his wife, Lord John Hay, the commodore on the Spanish coast, Sir James Craig, Margaret, Joanna, and James, and Sir George Warrender, a very pleasant party. It is another fine day, and I see the Ochills and Ben Ledi clear.

* Menzies of Pitfodels, a Catholic Aberdeenshire Laird,
converted Whitehouse into this Convent.

Saturday. I had a visit from Mr. Bryson the watchmaker, after breakfast, to talk about the School of Arts, and afterwards called on Professor Jameson. Then Lord and Lady Murray, Mrs. Primrose, Lord John Hay, and I went in the open carriage to the High Street, by the Castle Road, to see the great printing establishment of the School of Arts.

On Thursday the Gibson Craigs pressed me so kindly to come out to see them, that I gave up the Arthur-seat expedition, and went out this morning to breakfast. Lady Murray gave me her open carriage. It is a most glorious day, and the country is in the greatest beauty, and most especially Riccarton. Sir James took me into the garden before breakfast, and led me through a shrubbery to a small iron rail gate ; he took a well-polished small key out of his pocket, unlocked the gate, and after a short way through some shrubs and flowers, we came to a grass plot, surrounded with shrubs and flowers. " Here," he said, " Lady Craig lies," and pointed to a little hillock. It was a touching scene—she expressed a wish to be buried in the grounds, and there is, I think, a great satisfaction in having one they loved so much still kept, as it were, among them.

Margaret, Joanna, and Jemima were at home. Sir James was obliged to go to Edinburgh soon after breakfast, but they walked about the gardens and grounds with me till twelve, when I left them, very much gratified with my visit.

Love to dearest Mamma, and all around you.

Your affectionate father,

LEONARD HORNER.

From Mr. Edward Romilly.

Stratton Street, *June* 16*th*, 1841.

My Dear Mr. Horner,—I have been long intending to write to you, if only to express to you the pleasure I have had in reading that portion of your brother's papers which you have sent me, and to thank you for what I cannot but consider as a very flattering mark of your confidence. I wish 1 could make any adequate return for your kindness; but I fear that any opinion of mine can be but of little use to you, I may, however, at least tell you how interesting these papers were to me, and that I cannot doubt that a selection from them would be equally interesting to the public. There is, however, a peculiarity about your brother's character, as described in these papers, which although the most valuable, and certainly the most characteristic part of it, may possibly take away from its interest with the common herd of readers. I allude to his remarkable judgment, and to the steadiness of his conduct at the very outset of his career. He is a philosopher at nineteen, in thought, style, and character; a man of note after years of self-training, might well envy his moral and intellectual attainments at that early period of life. This very perfection, however, if it does not raise the suspicion that something remains untold, gives a somewhat artificial appearance to this part of the papers, and may prevent the common reader from sympathizing as fully with the character they describe, as might be wished. Most people when they are studying the character of a great man, especially his youthful days, please themselves with finding in it touches of nature, and even foibles, which they are conscious of having themselves experienced. The similarity, although in trivial matters, between their own character and their hero's, is flattering to self vanity; excites curiosity and commands attention; the absence of these traits, may therefore in some degree interfere with the *popularity* of the work, and limit the class of readers to those who know how to appreciate so great a peculiarity. Certainly, so remarkable an instance of precocity as that afforded by your brother's character, is very rare and very curious.

Many of the letters of his correspondents, especially those

of Jeffrey, are admirable; and the history of the origin of the *Edinburgh Review*, and of the editor's difficulties, are very interesting. Brougham's letters, too, are exceedingly characteristic. How far such letters can be made use of is a matter upon which I do not presume to give an opinion. We, fortunately, had few such difficulties in our case; for although your brother died before my father, my father's correspondents were of course of a much earlier date, and were almost all of them gone before we happened to be in a condition to publish at all. I foresee too that possibly a charge of irreligion, or want of religion, similar to that made against my father, may be made against your brother; and this charge would be strengthened by Sunday morning having been selected for his readings of Bacon, &c. All these things have, however, no doubt occurred to your own mind still more forcibly than to my own; and you will know how to deal with them much better than myself.

I certainly do regret that so few traits of his childhood have been preserved; but there is always some drawback in such papers; and one ought to be contented with what remains, without attaching too much importance to what has been lost.

Once more let me thank you for making me better acquainted with the history and character of one I have always so much admired, and whose memory will, I trust, be as much cherished by posterity as it ought to be. I do not know whether I ought to excuse myself for not having written to you before on this subject, or for having written at all. But in either case you will give me credit for the interest I feel in this matter, and for the sincerity with which I am,

Yours very truly and gratefully,

E. ROMILLY.

To his Wife.

Edinburgh, *20th June,* 1841.

I will wait till my return to answer E. Romilly's kind letter. His observations are very sensible. The points he touches upon I have, however, had for some time often in my mind; they are some of the chief difficulties of the undertaking. I thought Lord Lansdowne would have sent back the fourth and fifth volumes* by this time, and I hope he will not have left town before I return. Thank you for thinking of me in postponing your breakfast at Mr. Rogers', for I shall greatly enjoy going with you on the 28th, you had better write him a note to say how much pleasure I shall have in availing myself both of his invitation and of his offer to shew us Stafford House. You appear to have had a pleasant party at Lord Monteagle's, in Mansfield Street, I suppose? Have you seen or heard anything of the Hallams lately?

Robert Graham confirmed to me yesterday by his letter, what you had heard of Fox Maule leaving the Home Office. He is to be Vice-President of the Board of Trade, and the removal is set down to the strong side he has taken on the Scotch Church question, to the embarrassment of the Government, and which he will have nothing more to do with when he leaves the Home Office. Sheil is to be Judge Advocate, and Sir G. Grey, Chancellor of the Duchy of Lancaster, according to information here.

[In August Mr. Horner and part of his family, spent some weeks in Cornwall and he attended the British Association at Plymouth, and on his journey north in September he paid passing visits to Mr. and Mrs. Merivale at Barton Place, near Exeter, and to Mr. and Mrs. Sydney Smith at Combe Florey.]

* Of the MS. of letters copied.

To his Wife.

Barton Place, near Exeter, *1st September*, 1841.

I spent a couple of hours in Exeter, and then came here, where I met with a most hearty welcome from Mr. and Mrs. Merivale. He and I have been out ever since walking about his grounds. This is a far prettier place than I expected to find it, and in a letter I wrote to Mrs. Mallet this morning, I told her that she had never praised it half enough. It is really beautiful, and the view from my bed-room window this morning at six, when the sun was shining on the brilliant green fields, was charming. I am happy to see them so cheerful after their sad afflictions. Mrs. Merivale has been talking to me this morning about both her sons with great feeling and sense. The room I slept in, is furnished with Reginald's things, and they have been fitting up a bookcase in their library for Alexander's collection of the Classics, Charles came from London last night to pass a month with them. Merivale, to drive away afflicting thoughts, set actively to work at translations from the German, and he tells me that he has got far on in Schiller's lyrical pieces, which he means to complete and to publish ; they are verse translations of course.* He shewed me a most beautifully illustrated copy of the " Niebelungen Lied," recently published at Düsseldorf, and another still more beautiful work, a collection of tales in verse of a popular nature, illustrated with vignettes and embellished borders, also published at Düsseldorf.

Merivale and I walked above three hours in his own grounds yesterday, and although he has not more than a hundred and fifty acres, it is so diversified in lawns, fields, and woods, and has been so skilfully laid out, that he has above three miles of walk.

* Published in 1844.

To his Daughter.

Exeter, 3*rd September*, 1841.

MY DEAREST KATHARINE,—After finishing some letters yesterday, Mr. Merivale and I walked to the top of a hill behind
their house, called Mary Pole Brow, from which there is a
fine panoramic view. I saw all down the Exe to Exmouth
and Lympston, Powderham, Mamhead and Haldon Hill, which
we saw last year, and a sweep of the rich country around
with Hey Tor and Cawsand Beacon, the highest point of
Dartmoor, in the distance.

Mr. Merivale and I walked to Sir Stafford Northcote's, Pynes
House, about half a mile off. It is an old mansion, situated in
a noble park, with very fine trees of various kinds, and the
house is surrounded with a luxuriant mass of evergreens and
flowers. We found the Baronet at home, an infirm old man,
and he shewed us a few good pictures in the house, some by
Northcote, who claimed relationship with him. As we came
away, we met Mr. Henry Northcote, his eldest son, who
married a Miss Cockburn. I have been fortunate in having
had two very fine days to see Barton Place and the country
round, for to-day it has been raining ever since I got up.

Farewell, my dearest Katharine, and with love to your dear
Mamma and sisters.

I remain,

Your affectionate father,

LEONARD HORNER.

To his Daughter.

Manchester, *September* 8*th*, 1841.

MY DEAREST FRANCES,—I passed Friday with Dr. Miller,
in Exeter, an old friend of my own, and a still older and more
intimate friend of your poor uncle. He is a nephew of Dugald

Stewart, and we had much talk both of persons and things, in which we take a common interest. I went to the Sydney Smith's at Combe Florey, on Saturday, to dinner, and stayed with them till Monday morning. It is about seven miles from Taunton, and is the most perfect gem, in its way, you can imagine. Nature has done much for it, in the disposition of the ground, both in the place itself and in the surrounding country, and she continues to do a great deal by the luxuriant vegetation, the emerald turf and the noble trees; but she has been greatly assisted and decked out by the admirable skill and perfect taste of Mr. and Mrs. Smith, and especially the latter I believe, her skill and taste are equally shewn in the house itself, which has all the accompaniments of refined life. They were quite alone all the time, and both very kind and agreeable. He was very cheerful, and said some good things. On Sunday they gave me a horse, and I had a most delightful ride to the top, and along the ridge of the Quantock Hills. You *have heard* that thirty years ago your Mamma, Mary and I, spent some weeks at Alcombe, at one end of the range, and two years afterwards at Enmore at the other extremity, and cannot tell you with what pleasure I recognised numberless spots with which my hammer, and our expeditions, made me familiar. It was like meeting old friends after an absence of thirty years, and these friends unchanged. It was a glorious day and I had a vast panorama. I suppose there is not in the world such an expanse of richly cultivated land, and that too with the addition of a fine landscape—hills, mountains, woods, pastures, and the Bristol Channel studded with islands and ships. But I did not take my ride until after church, where I heard Mr. Smith preach a most admirable sermon from his own pulpit, in which he showed the error of those who make too great a show of piety, and make religion morose, judging harshly of those who seem to them less holy than themselves. His text was taken from the parable of the Pharisee and the Publican.   I left Combe Florey after breakfast on Monday, and got by coach to Bridgewater, by mid-day; there I got upon the Great Western Railway, by far the best I have yet travelled upon, and at six was at Cirencester, 92 miles. I found a coach to take me to Cheltenham, 15 miles, and there I found a railway which conveyed me to Birmingham, and at seven I arrived here.

Give my love to dearest Susan,

Your affectionate father,

LEONARD HORNER.

To Charles Lyell, Esq.

Liverpool, *October* 3rd, 1841.

MY DEAR CHARLES,—I have read half through the second volume of the "Elements," and continue to be much pleased and instructed.   I have made several notes for your consideration when you come to a third edition, which I hope will be a part of your autumn work next year.

I found John Prévost here, and dined with him at Henry Romilly's.   Melly is gone with his family to pass the winter at Naples and in Sicily.   Lucky fellow! what would I not give to be in his place!   Romilly got me an order to see the " Columbia," for I was curious to see one of those celebrated steamers.   It is certainly a noble vessel as far as I can judge of such matters.

The stagnation in the manufacturing districts continues, and there are apprehensions stated repeatedly that the suffering among the work people will be great this winter. The Corn Law agitators are in full activity, and meetings are being held all over the country to petition the Queen not to prorogue Parliament until it has taken into consideration the distresses of her people, and especially the question of the Corn Law as one great source of them.   But the Duke of Wellington and Peel, strong in their Parliamentary majority, care not a rush for these meetings, and Parliament is to be prorogued on Thursday next, it is said, without any expectation of its meeting again before February. I can well understand that the Government require time to consider the details of any measure, but they remain obstinately silent whether they will make any alteration at all, either in the Corn Laws or any other part of our commercial code, and that I think they are justly condemned for. I saw Tuffnell, the Assistant Poor Law Commissioner, at Rochdale a few days ago. He is the chief associate with Kay in the Battersea School. He says that from what Kay has seen of the new members of the Education Committee of the Privy council, he augurs favourably, and expects that the Government will seek popularity by doing something considerable for the national education. I hope it may be so.

Give my kindest love to my dear Mary, and believe me, my dear Charles,

Your affectionate father-in-law,
LEONARD HORNER.

To his Wife.

Manchester, 21st *October*, 1841.

. . Tell Mallet that I saw Smyth in Leeds, and thought him looking *fresh* and well, and he was in good spirits. But I only saw him for half an hour, and should have found half a day too short a visit, whatever he might have thought.

You ask me about the distress here. It is certainly very great, but at present more among the masters, I believe, than among the workpeople. There are a great many of these out of employment, and I am told that, owing to the diminished wages of many, that they are sorely pinched, and many getting into debt to the shopkeepers. The pressure is greater in some places than in others. In Stockport it is said to be very bad, but that is not in my district. In Saddleworth they are very

badly off.   A better sort of workman, a manager of a small
mill, but not paid by weekly wages, told me they had not had
on an average, two days' work in the week for three months.
I visited his cottage and saw his wife ; they have nine children,
and he told me he had got £12 in debt.   I was much struck
with the patient way in which they spoke of their privations,
and the pious expressions of hope that better times would
come, no angry expressions, no rude complaints, and they
might have been uttered and excused.   An extensive mill-owner
in Delph told me that a young man left his employment a
week ago and he had nine applications for the situation, all
from young, stout men between 20 and 25 years of age, who
were out of work, and the place was only 12s. a week.

Mills for which £72,000 was offered in January, 1839, were
sold four months ago for £36,000, and last month machinery
which in ordinary times would have fetched £20,000 was sold
for £7,500.   These are facts from which a judgment may be
formed of the state of things, and I have mentioned them in
my report.

———

To his Daughter.

Manchester, *October 24th*, 1841.

MY DEAREST JOANNA,—Since I finished Hamilton's " Letters
on America," I have resisted getting another light book, and
I have taken up the "Principles of Geology," having only
read part of the sixth edition.   I began at the beginning and
am so absorbed in the interest of it, that I do not think I
shall take up another book until I have gone through the
three volumes.   I rise from each perusal with fresh delight,
and more intense interest in the sublimity of the subject,
which by carrying us back not only into infinite distance, but
by progressive stages backwards, and unfolding the beautiful
operation of eternal laws fixed by the great Creator, raises

one's adoration of His infinite power and wisdom, and gratitude that he has endowed us with faculties capable of comprehending and enjoying such sublime works. The " Principles " ought to be read and studied by all of you, there is not a page that is not quite within the reach of your mind, without more scientific preparation, with such explanations as I can give you ; and I am sure the perusal would have the effect of enlarging your minds, and giving you juster views of the world you live on, without the necessity of your knowing one stone or one fossil from another. It is the great general truths connected with the history, past and present, of the globe, that I allude to, when I speak of the beneficial effects the book would have in enlarging your minds. My love to dearest Mamma and your sisters.

Ever dearest Joanna,
Your affectionate father,
LEONARD HORNER.

To Charles Lyell, Esq.

Manchester, November 2nd, 1841.

MY DEAR CHARLES,—I had the satisfaction of hearing this morning of your return to Boston.

I had a letter from Susan dated the 29th and written from Tarbert on Loch Lomond, where they were with Lady Murray, on their way to Edinburgh. They have been passing a very happy week at Strachur, diversified by a ball at Inverary Castle, and another day spent there with the Murrays. Lord Murray said that he should pay a visit to Sir Archibald Campbell near Glasgow, on his way to Edinburgh, which I have no doubt he did on purpose that Frances and Susan might go with Lady Murray in her carriage to Edinburgh ; they were to remain with Lady Murray till to-morrow, and then go to the Cockburns.

I shall meet them at Lancaster and I expect we shall leave this for home on the 27th.

Your mother-in-law says in her letter to-day, that Fitton had been calling on her, very happy to be again in his house in Harley Street. He told her of letters from Murchison, who has found the Silurian beds on both sides of the Ural Chain. He is on his way home. He and de Verneuil were upset in a boat on a stream, lost all their portmanteaux, and how valuable their contents were, does not appear. Roderick held his note book and *hammer* above water, and de Verneuil his fowling piece.

I have had Hallam here since Friday. He said in a letter from Bowood that he would come here if I was here, as he had a desire to see the manufactures under my guidance, and I have been devoting myself to him. He appears much interested with what he has seen, and particularly with hearing, from authentic sources, details respecting the present unparalleled and severe state of distress. There are very serious alarms for the winter, and the sufferings, both among masters and their workpeople, will be very great.

I took Hallam to the Phillip's at Barlow Hall, and we stayed till Monday. He is very well, and in good spirits. You will be much pleased to hear of Whewell's being made Master of Trinity College, on the resignation of Wordsworth. Hallam thinks it a capital appointment, that he will do much good to the College. At all events it is not probable that he will give any countenance to the bigoted system that is now gaining ground at Oxford, in the discouragement of the study of the physical sciences. I had some talk with Hallam about it on Sunday, who deplores and condemns their proceedings in the strongest way. You know Hallam's general opinion, and that he is an Oxford man, and he used these words to me " I should as soon think of sending my son to Maynooth as to Oxford at this time." Whewell within ten days got both his wife and his mastership. The accounts you give, both in your own letters and Mary's, of the pleasure and instruction

you are receiving, is very delightful, and above all that you
are keeping your health so well.

Your accounts of the coal-fields you have been seeing are
very interesting, and Mary says you have sent over a paper
on some of them for the Geological Society. I am sorry that
I shall not be in town to hear it.

To Charles Lyell, Esq.

London, *2nd December*, 1841.

MY DEAR CHARLES,—On returning from the Geological
Society last night, I had the great pleasure of hearing dearest
Mary's letter of the 14th ult. At the Council, I had the
satisfaction of seeing Darwin again in his place and looking
well. He tried the last evening meeting, but found it too
much, but I hope before the end of the season he will find
himself equal to that also. I hail Darwin's recovery as a vast
gain to science. There was nothing very particular trans-
acted at the Council ; we had a good club, but small, the only
stranger was Mr. Logan of Swansea, arrived a week ago from
the United States. He told me that he had seen you and
Mary at New York in August.

The evening began with an account of a great earthquake
that took place at Terceira, one of the Azores, last June,
which threw down the whole town of Praya, consisting of five
hundred houses. Some facts that were mentioned are interesting, particularly a great rent of a mile long from the shore
inland, and the rising up in the sea of a shoal, which shortly
afterwards was washed away. It was stated that several such
shoals exist, raised within a short period, and that the navi-
gation is thereby rendered very dangerous. The paper was
sent to us by Lord Aberdeen, Secretary for Foreign Affairs.

Next we had an account by the Rev. Mr. Everest (who was
present) of a journey from Delhi to a place considerably

within the range of the Himalayas. He states that as you
approach the central range, the same rocks assume a metamorphic character, and he describes some remarkable rocks of
granite intersected by granite veins. But the most interesting
part of the paper was the announcement of vast numbers of
monkeys living at an elevation of 8,000 feet above the sea. I
sat next him, and he told me that he lived in that spot for a
year, that in the month of February, while the country far
and wide was covered with deep snow, he has seen groups of
the monkeys sitting on the pine trees picking the seeds out of
the cones, and feeding upon them. He said they were surrounded with them. Owen was much struck with this
announcement. Everest also stated that he saw the skin of
a leopard that had just been shot at an elevation of 11,000
feet, and that they are frequently met with, as they follow the
wild goats to those heights. He told me that he had seen the
Bengal tiger amidst the snow at 8000 feet elevation, that they
do not go higher he believes only because the mountains are
too precipitous for their heavy bodies to climb, whereas the
lighter and more agile leopard finds less difficulty. He says
that the species of monkey is identical with that living in the
plains.

We had then a paper from Owen on fossil *marine* turtles,
found in the London clay of Sheppy and Harwich. Hitherto
it had been supposed that the Chelonian remains found in
that deposit, belonged exclusively to the Emys, but on
examining specimens in Bowerbanks collection, and others
sent to him by Sedgwick, Sir P. Egerton, and Dr. Dickson of
Worthing, he has made out six distinct species of marine
turtles. He told us that there are only six living species
known, and that never more than two have been found
together in one region, whereas here, in this confined spot,
six have been found. There is, however, this remarkable
anomaly, that whereas all the living species attain a consider-

able size, none of those found in the London clay exceed ten inches in length, hence he supposes that they lived in brackish water in an estuary. He gave us a most lucid description of the specimens on the table. You will see by the newspaper that your friend Sir Edmund Head is appointed the Poor Law Commissioner in the room of John Lefevre, who resigned. You will see also the death of Chantrey. He had a small dinner party, and in the evening fell back and expired in the arms of Stokes.

I shall be very busy for the next two months, partly with my own official matters, and with the Children's Employment Commission, but I shall reserve some time for geology, and hope to be able to make out a little notice on Cawsand Bay. I shall also be very busy with my brother's papers, and expect to have the transcript of the whole completed in a month hence.

Mary asked me some time ago the origin of the term cannel coal. I believe the etymology to be nothing more profound than the Lancashire pronunciation of candle, as the Wigan people burn it to give light in their cottages.

CHAPTER III.

1842—1843.

To Charles Lyell, Esq.

London, 24*th February*, 1842.

MY DEAR CHARLES,—Last Friday we had our Anniversary
at the Geological Society, there was a full attendance—the
Wollaston medal was awarded to Von Buch, but Bunsen was
unwell and could not attend to receive it. Murchison's
address was excellent, very comprehensive in his view of what
has been doing in the last year, and with a great deal of
valuable matter. Greenough said that it had never been
surpassed by former addresses. We had a good party at the
dinner, about ninety, and some excellent speaking. Murchison
had Baron Brunow, the Russian Ambassador, on his right, the
Duke of Richmond on his left, and farther on the right Lord
Lansdowne and Lord Sandon, on the left Lord Enniskillen
and Sir John Johnstone. I was well placed between Hallam
and Symonds.

Last night we had an admirable paper by Owen. A
description of the fossil skeleton now exhibiting by Herr Koch
at the Egyptian Hall, the man you saw at New York; I will
try to give you an outline of the paper. This great Missourian
quadruped, as Owen termed it, has been made a monster both
in name and in putting together by Koch—he has lengthened
the back bone by intervertebral substances, and by additional
vertebræ by a wrong position of the scapulæ, he has raised
up a ridge on the back, and he has added a lower jaw which
did not belong to the individual, and the penetrating skill of
Owen discovered that he had unwittingly suspended a *female*

under jaw to a male. When Owen detected this, he admitted that the lower jaw of the individual found in the pit, and which he represents as having been bogged in a hole, was wanting. Then he has placed the tusks so that they extend horizontally, whereas their true position is with their curvature upwards—Owen states that the roots of the tusks and the sockets are quite cylindrical, so that when the ligatures or soft animal substance which held them fast was destroyed, they might be easily turned round in any direction, and so it is very possible that what the man says may be true, that he found them extending horizontally. Owen considers it a male Mastodon Giganteum, and he went on to shew that that genus, and the Tetracaulodon of Godman, are one and the same. He says that hitherto, naturalists have been led into errors, in considering individuals as belonging to distinct genera and species by the state of the tusks. In the male and female, there are the rudiments of the same tusks, *but they are never developed in the female*, and he points out the analogy in this respect with the Dugong and Narwal. It is particularly striking in the latter, for the long projecting straight ivory twisted spear, which gives it the name of Unicorn, is nothing more than a tooth, an elongated tusk, and what is curious, the corresponding tusk in the other ramus is never developed, but continues in its rudimentary state. He considers the Tetracaulodon to be a Masterdon in an immature state. There is one short straight tusk in the lower jaw of the Mastodon, the use of which is very obscure. After he had finished his paper, he said, he received Dr. Hayes' of Philadelphia (I think), paper on the Tetracaulodon, but after going over it carefully he saw no ground to alter the conclusion he had come to before, *viz.*, that the Mastodon Giganteum and Tetracaulodon belong to the same genus. The paper was much eulogized by Buckland. Charlesworth, who stated that he had recently been examining specimens in museums at New York, Philadelphia, and Baltimore, spoke, and Owen told us afterwards that he had spoken well. After the paper we had a very masterly verbal elucidation of the case by Owen, referring to an admirable drawing made under his direction by our sub-curator, Woodward, about seven feet long. After the meeting, we adjourned to Lord Enniskillen's ; Owen, Clift, Buckland, Fitton, Major Clarke, and Edward Bunbury and his brother, and we had Crustacæ, carbonized fragments of Costoe of a mammal (probably Bos-broiled boniensis), and much smoke and merriment. Poor Egerton was attending at St. Stephen's, in the expectation of a division on the Corn-Laws.

Manchester, 2nd March, 1842.—I am settled here as head-quarters for the next six weeks, unless I am sent for to London by Sir James Graham, about the Factory Bill ; but I have no expectation of that, for he has said that they will not bring it forward till after Easter.

Anthony Symonds has just made a very careful levelling from the Mediterranean to the shores of the Dead Sea, with an excellent theodolite, and after going twice over his calculations, he makes the level of the Dead Sea more than one thousand three hundred feet below the Mediterranean. I do not see that geologically there is anything to make his measurements questionable, because of the great depression, and that it may be as well one thousand three hundred as twenty feet, but it is a curious fact in physical geography.

I think I mentioned to you that I was put upon a Committee of the British Association for the purpose of taking steps to record earthquakes that occur in England, with Buckland and Wheatstone. We had a meeting ten days ago, and we have set Wheatstone to work, to contrive an instrument that shall be accurate, ready for use, and cheap. He described a contrivance to me on Saturday, which promises to be very effective. He has a most extraordinary fertility of invention,

I wish he had the faculty of finding money as readily. I fear
that his electric telegraph will never bring him a shilling.
Edward Bunbury has been doing duty as an Examiner at
Trinity. He says that Whewell's appointment is very
agreeable, that he is making his house very agreeable, that
hitherto heads of houses have visited none under the
Academical grade of the head of a college (a most absurd
custom by the way), but Whewell has broken through
it, and has announced that his house will be open to those he
likes.

Ever yours,

LEONARD HORNER.

To his Wife.

Manchester, 7*th March,* 1842.

I have been solely engaged since I wrote yesterday in my
leisure hours with the memoirs, and a very little reading. It
is a great satisfaction that after the trouble I took in preparing
my last report, it has not only not been passed over unnoticed,
but has been commented upon so often in Parliament.
Charles Buller's notice has been the fullest I have seen. It
has also given great satisfaction to the mill-owners, several
have thanked me for it, and some of the most acute have
said that they have read it with great care, and have not
found any errors in it.

So you see that we are to have a property tax. I like
Peel's speech, he appears to me to take the bull by
the horns. I have not yet seen any of the other speeches. I
am curious to know what is thought of it by those you
see. Until we make up our minds to give up a great part of
our over-grown Colonial possessions, which are a dreadful
drain upon our finances, we shall never be able to make any
considerable reduction in our expenses. I go to-morrow to

Rochdale, but return at night, the railroads all around Manchester have greatly facilitated my movements.

To his Daughter.

Rectory, Bury, 3*rd April,* 1842.

MY DEAREST JOANNA,—I came here yesterday to dinner. Mr. and Mrs. Hornby are alone, for Miss Hornby is at Croxteth Park, on a visit to her cousin, Lady Sefton. I have just come from church, where Mr. Hornby gave us a good sermon. It is a bright, clear day, but the wind is high and blows as cold as if it came from a glacier. I see from my window the house where Sir Robert Peel was born, where his father had print works, and laid the foundation of his vast fortune and his son's great station. His father was a shrewd, clever tradesman, and lived in the days of enormous profits. As an instance of the contrast of profits of those days and the present, I was told yesterday of his having written to one of his travellers to return, because he could not make more than a guinea profit on each piece of calico he sold; whereas now they think themselves very well off if they get one shilling of profit on each piece.

From Lord Jeffrey.

Craigcrook, *July* 26*th,* 1842.

MY DEAR HORNER,—Your kind letter came when I was in the hurry of writing up my arrears of *judgments,* and has been a little forgotten, I fear, since.

I am much obliged to you for your congratulations and enquiries about my health, which is certainly so far mended as to have enabled me to go through the whole of my summer sessions without having been a day out of court. But I have no vigour yet to boast of, and I am afraid must lay my account with being somewhere about *seven years older* than I was, little more than one calendar year ago!

And now as to your intended publication. I am sure I need not say that I shall look for its appearance with a very peculiar interest, and be both proud and happy to contribute any little thing I can, to its completeness.

As to my letters, if you think they can help at all to illustrate any part of my dear and early friend's character or pursuits, I can have no other motive for withholding them than may arise from feeling that they may be thought *too* unsuitable with the gravity and propriety that *should* belong to my present judicial character. However, I shall be better able to judge of this when I see them, for my memory is entirely a blank, either as to their subjects or general tone, and indeed I have so much reliance on your sound discretion, as not to have much apprehension that *you* could propose to print anything which *I* should wish to suppress. As to the sketch of character which you suggest, I feel much more hesitation. My impaired health and strength naturally increase this reluctance, and though I do not *absolutely* decline your very flattering request, I must beg to be allowed to gather together my recollections, and consider what I may be able to make of them, before I give any definite answer.

Mrs. Jeffrey desires me to send her most kind remembrances to you, and such of your family as may be with you, and joins with me in hoping that the season may not pass over without our seeing some of you in Scotland.

Cockburn has had a short, but sharp, attack of illness, but is well again, and walked with me half an hour to-day in his square.

With all good wishes,

Believe me, ever very faithfully yours,

F. JEFFREY.

To his Daughter.

Bedford Place, *Tuesday, August 2nd,* 1842.

MY DEAREST MARY,—It was a great comfort to me to receive
your letter the day before yesterday, as your illness at
Montreal made me more than usually desirous to hear again.
Thank God that you were both quite well, and I shall almost
count the days until you arrive. If you have an average passage
you should be at Liverpool by the 29th, so we shall be looking
out for you. The result of the consultation with Sir J. Clark
and Mr. Quain yesterday was, that were I to attempt to go
through an ordinary day's work of inspection, I should probably have a serious relapse of inflammation and must not
think of moving into Lancashire until I can go freely about here.
I have been to-day to the Home Office to Mr. Manners Sutton,
than whom no one could be more kind, and he at once put the
matter on such a footing as to make me quite at my ease.
My doctors advised my going back immediately to the country,
so your Mamma and I will return to Waverley to-morrow.

Edward Bunbury sat two hours with us last night. He sets
off on Friday for Guernsey to visit Col. Wm. Napier, who is
now Governor there; thence he crosses to St. Malo, and
proceeds at once to the Pyrenees. After seeing them well in
several parts he means to go into Spain, and from Gibraltar
to Malta, thence to Rome where he will pass two months. He
does not expect to be in London again before the end of May.
We shall miss him very much. Darwin continues well. You
probably heard that poor Dr. Arnold of Rugby died about six
weeks ago; the election of his successor took place last Thursday, there were many competitors of high claims, men with
great University reputations, and who do you think was the
successful one? *Archie Tait,* our dux of the Academy, who
has since continued in the same career of distinction, having
worked up to be senior tutor of Balliol, and University
Examiner. Is not this a great feather in the cap of the

Academy? I have completed the memoir of your uncle, and am ready to go to press with the whole, as soon as Murray is disposed to begin. I got, a few days ago, three valuable letters of your uncle's with a kind note from Lady Grenville; they were addressed to Lord G. I applied to her through Lord Lansdowne. She expects to be able to send me some more.

I hope this will find you within a day of your embarkation at Halifax, and that you will have a pleasant and prosperous voyage. My kindest love to Charles.

Your most affectionate father,
LEONARD HORNER.

P.S.—The Darwins have just been here, and are both well; they send their kindest regards, and bid me say how happy they will be to see you both. In a note to your Mamma last night he said, " I long to see my patron and master again," meaning Charles. They have seen a house in Kent beyond Bromley, about sixteen miles from London, which they have serious thoughts of taking.

From Lord Jeffrey.

Craigcrook, 11*th August,* 1842.

MY DEAR HORNER,—I now return you the letters, which if you continue to think worth printing, I do not wish to withdraw. You will observe that I have scored out the name of *Brougham* in the letter of the 6th December, 1808. There is no spirit so irritable that yet walks the earth as his. By the way, you must come across him, I fear, in a far more serious and painful way, in your account of your brother's early *liasons* with him —and his subsequent alienation—though it is possible that you may have thought it best to say nothing of the latter, yet it weighed so long and heavily on the affectionate and generous heart of your brother, that it must have had a prominent place in his confidential correspondence. I know that *I* had several noble and touching letters in relation to it.

I am very sorry to find that your malady still hangs about you, and interferes with the discharge of your duties. But the first of all duties is to the doctor, and that duly paid, will soon make all the others easy. You have every advantage, I am sure, of weather.

For my part, I hope I continue to improve a little, and at all events have no suffering to justify complaint or repining.

Will you not run down to see us before the end of the season? I shall be here I think pretty steadily till the courts meet on 1st November, and too happy to see you. With kindest regards to Mrs. Horner, and all your house.

Ever faithfully yours,

F. JEFFREY.

To his Wife.

Lea Hurst, 13*th September*, 1842.

I remained at Rugby from ten minutes past one to three, but I had a book, so did not mind. The station is some way from the town, and I therefore did not go in search of Tait. I got to Derby at half-past five, and started again at six, and in half-an-hour stopped at the Amber Gate Station, where I learned that Mrs. Nightingale had been two hours before. I got a phaeton, which brought me here, four miles from the station, by a quarter past seven, when I found they had just sat down to dinner, having given me up. They gave me a hearty welcome. No stranger in the house but myself. They are all well, and both the girls looking very well. From Amber-Gate the road is very pretty, but there was scarcely light enough to see it. When I awoke this morning, I looked out on a most beautiful view, a rich wooded valley with hills and finely broken ground on each side. It is a small but very

pretty house inside, and as far as my short acquaintance with
it goes, a most enviable spot to possess. Mrs. Nightingale is
going to take me a drive to Matlock.

From Lord Cockburn.

7th November, 1842

MY DEAR HORNER,—I have such sympathy with your
undertaking, that if I could do you any good, I would. But I
absolutely cannot. You must observe that I was *not* at the
class with Frank, except *one* year, being my first year, and
his last with the Rector. And even during that single year, I
had no acquaintance with him beyond sitting at opposite
extremities in the same room. What I said to Mary was,
merely, that I then knew him by sight, and used even then,
to be struck with the air of thought and gentleness with
which he used to make his retreat from the boys, whenever
the class was dismissed, and walk home, with his hands
closed, one over the other, across, and in *front* of him. I see
the figure yet. But what could you make of this? It is
nothing. It does not admit of being extended, and is not
worth extending. Besides, I have been told that I must be
dreaming, for he was a merry and not unapt to be an idle
lad, which, however, nothing will make me believe.

The most striking occurrence in his High School history, I
should suppose, must have been in its last day; when he, in
the name of the class, presented Adam with a book, which
they had subscribed for, in testimony of their respect and
gratitude. This was probably Frank's first public speech. It
was in Latin—his own composition—well thought, well composed, and well spoken. I see this scene too. But I was a
dolt then, and could not appreciate it. But surely Pillans
must be able to tell you all about it.

What can I say more? Ever,

H. COCKBURN.

From Mr. Hallam.

Wilton Crescent, *November 14th,* 1842.

MY DEAR HORNER,—I think that it will be quite right to
have a short biographical notice of Seymour,* and will
prepare one, as you desire, when you return to town. The
printing will, I suppose, not begin till that time.

Julia is at Clifton or at Clevedon for the present. Next
week I meet her on a visit to the Nightingales at Embley,
and she will return with me to town about the 24th. After
that time I do not look forward to any migration.

Your account of the state of things at Manchester is bad,
though not much worse than I expected. The American tariff
seems a severe blow, yet it must be seen that a country
having the raw material at home, and every advantage in fact,
&c., to manufacture it, must before long supply itself, even
without a protecting scale of duties. Capital for erecting
machinery is now wanting, but it will gradually come to
America by their own industry, or be transported from
Europe. Our hopes of clothing the whole of mankind with
our cottons, and cutting all their meat with our knives, have
been too ambitious.

I am to meet the Lyells to-day at the E. Romilly's—he
seems none the worse for his expedition.

Poor Lady Callcott, I fear, is verging fast to her end. I
saw her yesterday much feebler than before, and the medical
men do not expect her to rally.

Yours very truly,

H. HALLAM.

———

From Mr. Hallam.

London, W.C., *22nd November,* 1842.

MY DEAR HORNER,—I have been struck back by your letter,
requesting a biographical notice of Lord Webb Seymour to be

* Lord Webb Seymour.

inserted immediately. You are aware that I am going to Embley to-morrow, and have not yet prepared a word of what I promised you. But as necessity makes men desperate, I shall tell you what has occurred to me since your letter arrived. I have between twenty and thirty letters of Seymour's, many of which are very illustrative of his character, and may perhaps, for I have not read many now, be of some use to your memoir. He was also so valuable a man, and so little known, in proportion to his deserts, that a mere note seems to me a very inadequate mode of bringing him before the public. I should prefer an Appendix either to your first or second volume, probably the former, wherein somewhat more space may be allotted to his memory.

I find in a letter to me dated Edinburgh, 10th November, 1799, soon after his arrival there, the following passage. "Since my arrival I have rather been dabbling in different branches of science, than attending seriously to any. I have been engaged in some stout metaphysical discussions on time and space, cause and effect, and such light topics, with a friend whose name is Horner. From these we have derived no great advantage, except that we may suppose our faculties to have been strengthened by the exercise."

Could you not insert this in a note, when your brother first mentions Seymour, and then, after stating in two or three lines who he was, his virtues and abilities (in a general way), and his friendship for several distinguished persons, especially Playfair, refer to an Appendix for a fuller account.

You will be concerned to hear that I have this morning received an announcement of Lady Callott's death. It took place about last midnight without a struggle. I was there yesterday, but she was not in a state to see me. I saw him, who seemed quite prepared for the worst, yet Chambers doubted whether she might not last longer. Though he will bear the *privation* rather than the immediate loss very well,

I dread the effect upon a man weakened by his own illnesses, and obliged to lead a sedentary and solitary life; the continued watching she has required, though it certainly impaired his health, gave him an occupation which he will now want.

We go to-morrow to Embley, I am not sure whether I may not pay another visit. You will, I suppose, return soon to town, we can then talk over the matter of Seymour's biography, unluckily, I saw little of him, as you know, for many years.

Remember me to Mrs. Horner, Julia comes back to-day to go with me.

Yours very truly,
H. H.

From Lord Jeffrey.

Edinburgh, 10*th February*, 1843.

MY DEAR HORNER,—I am more mortified and dissatisfied with myself than I can expect you to believe, at my poor failure in the task you did me the honour of assigning me, I was not, however, aware till I received your letter of 28th January that it must be completed or renounced at so early a period, and, reckoning on a longer time of preparation, neglected beginning, or even finally making up my mind, till it was too late to think of the attempt. As I never had any notion, however, that I could do anything that would please either you or myself, there is the less cause to regret this result, and I really believe that your disappointment, from the experiment not having been made, is lighter than you probably would have had to encounter from its mis-carriage.

I forgot whether I told you myself, or only desired Murray to say, that I had read the sheets you sent me, with the deepest and most tender interest, and had been indebted to them for the revival of many very pleasing though now

somewhat sad recollections. I shall return them as you
request, either by to-night's or to-morrow's mail.

Murray is quite well again, and all the better, indeed, for
the discipline he has gone through. I have scrambled through
the Session, thus far, very tolerably, never having been a day
out of court, and feeling *at least* as strong and hearty as at its
commencement.

We shall be up in your latitudes, I hope, before the end of
March, but I imagine mostly at Haileybury, such a quiet
retreat being *safer* for my poor virtue, than the stir and
seductions of London. But we must contrive, I trust, to
meet notwithstanding.

With kindest remembrances to Mrs. Horner and your
daughters.

Ever your faithful, though feeble and penitent,

F. JEFFREY.

[The Memoir of Francis Horner, edited by his brother, was
published in March, 1843, and Mr. Horner received the
following letters from his friends soon after.]

From Mr. Hallam.

Wilton Crescent, *March* 10*th*, 1843.

MY DEAR HORNER,—I have run over, in parts, your two
volumes, for which I thank you. More and more am I confirmed in the opinion that you have erected a monument to
your brother's memory which will not perish. It will become,
if I mistake not, a popular book, not perhaps to the extent
of the memoirs of Romilly, but so as to convince you that
you have not failed of your aim in the publication. As far as
I can judge, from only reading in detached portions, you have
printed no letter that should have been kept back. I am half
sorry that you applied to Sydney Smith, whose letter is in
baddish taste, and contains very little.

Lyell's lecture on Tuesday was very interesting, I cannot go to-day, which I much regret.

Truly yours,
H. HALLAM.

From Mrs. Sydney Smith to Mrs. Horner.

Saturday, March 11th, 1843.

MY DEAR MRS. HORNER,—I had intended myself the pleasure of calling upon you yesterday, to have thanked you and Mr. Horner for the *most* valuable book in our possession. A severe cold caught at Mrs. Grote's the night before, has obliged me to draw into my shell, and shut my doors, that I mav not be tempted to speak. How very much the world is indebted to your husband for laying before it such a bright image of perfect purity, of kindly devotional affections, of high and unbending principle.

All this is not *new* to *me*, but I joy to think, as I did when Sir Samuel Romilly's memoirs were laid before the public, that such a bright example of unerring virtue is given to the world. These are the moral land marks, that will guide many others aright, or at least recall them if they do wander.

Yours, my dear Mrs. Horner, very sincerely,
CATHARINE AMELIA SMITH.

From Lord Murray.

11, Great Stuart Street, Edinburgh, *March 12th,* 1843.

MY DEAR HORNER,—I feel most deeply your kindness, which gives me great gratification, but that is not without distressing reflections.

Nothing can in itself be more pleasing to me than to be any way associated with my dearest and greatest friend. I have felt and do still feel a most feverish anxiety for the success and reputation of your publication, and I have an apprehension that the dedication to me may not promote either, but the reverse, as far as such a matter goes. I say this to you in confidence, as I conceal nothing that is uppermost in my mind on receipt of your letter. It makes me only more sensible of the kindness and partiality with which you regard your sincere friend,

J. A. MURRAY.

From Lord Cockburn.

Edinburgh, 14th March, 1843.

MY DEAR HORNER,—I got your two vols. late last night; and have as yet only had time to run over several pages in a loose desultory way, before setting to next week, at Bonaly, to a regular perusal. But I have seen enough to satisfy me of the propriety of what you have done. You have raised a lofty, because it is a just and pure, monument over your brother's ashes. No young man, improving himself in solitude, either for science, for public life, or for virtue, can resist the inspiration of such a record. What a train of events, of individuals of past years, does it recall to me! I presume that you have a good reason for omitting from the " Tributes to the memory," the character given of Frank by Brougham in the preface to one of his speeches, which is perhaps the best and most discriminating panegyric on your brother that exists.* But after all, it may be in the volumes though I have not yet observed it. There is so little about him in the work, that you may lay your account with his displeasure, which, however, seems less offensive than his friendship. He was left out of Romilly's memoirs in the same way. So far as I can yet judge, these memoirs of your brother are better in every way than Romilly's.

Ever,

H. COCKBURN.

* See Journal of Henry Cockburn, Vol. I, page 192.

*From Mr. William Erskine.**

Edinburgh, 18*th March*, 1843.

MY DEAR SIR,—I cannot tell you with what intense interest
I perused the Memoirs and Correspondence of your brother,
with a copy of which you have so kindly favoured me, and
which I most deeply prize, as a testimony of your regard and
of his. They are a most faithful portraiture of the original, and
reflect his mind in every page. From his first efforts in self-
education, a study which occupied to an uncommon degree the
earlier portion of his life, down to the full completion and
success of his efforts, as exhibited in the later period of his
brief career, all is most interesting. The mixture of soft and
tender affection for his family and friends, with inflexible
firmness in the discharge of every duty, public or private, is
very delightful; and the high eminence which he attained in
public life, the admiration of every class of men, without the
shadow of sacrifice to the prejudices of the high or the low of
any degree, places his character more among the memorable
men of Grecian times than of these our days, and yet the
universal admiration that he commanded is honourable to
our times also, and shews that the feeling of virtue is still
deep-seated among us. I know no volumes better fitted to be
a study to the young. They naturally take their place beside
the memoirs of his friend Romilly. You judge rightly that I
perused in particular with profound interest, the earlier
portion of his life, and that the narrative recalled many a
scene to my recollection of him and of our youthful associations. Let me thank you, my dear Sir, for the very great
kindness with which you speak of me, the humblest but not
the least sincere of his friends.

Believe me always, my dear Sir,

Most sincerely yours,

WM. ERSKINE.

° Author of the Memoir of the Emperor Baber, &c.

From Mr. James Reddie.

Glasgow, *22nd March*, 1843.

MY DEAR SIR,—I have delayed answering your very kind
note, until I could say I had received (yesterday) your
invaluable present. But I had previously looked into your,
to me, most interesting work, and again wept over the
" reminiscences " of the warmth of early friendship. I must
not, however, yield to the weakness of the *now* OLD MAN, and
will only add, with regard to myself, how highly I feel
gratified by your brother's approbation, although conscious he
infinitely *over*-estimated me, from the kind partiality of his
benevolent heart.

Although I have not yet had leisure to go regularly through
your volumes, I have read enough to be delighted with the
chaste elegancy and correct taste of your narrative, when
you do allow yourself to come forward. And I rejoice to
observe your brother's own diary, combined with his letters,
have enabled you to give so true and complete a picture
of the high-toned feeling and enlarged views of his enlightened
mind, a mind evidently destined for even more elevated
occupation in a wider sphere.

With kindest regards and best wishes,

I am, my dear Sir,

Yours most faithfully.

J. REDDIE.

From Lord Cockburn.

Bonaly, *24th March*, 1843.

MY DEAR HORNER,—I have now read the whole book, with
the greatest attention, and the greatest delight. So far as I
hear, or over-hear, everyone in this region is pleased. For
myself, I must have ceased to have either memory or heart, if
I could follow a progress from its opening to its close, which

I watched with such interest as it advanced, witho ut the deepest sympathy. It recalls my own whole life, so far as h is lasted, and though nothing could have increased my regard and admiration of him, this disclosure of himself has given me better *reasons* than I had for my reverence. The rearing of such a monument over the ashes of such a brother, ought to gratify you all the rest of your life.

25th March.—And how much I have been charmed with every word from, or about Seymour ! His letters have raised him greatly in my view for talent—some of these are really excellent, such as his exposition* of the causes and effects of the errors of coteries of party-men in London. Hallam's sketch of his life and character is be autiful, affectionate and just.

I wish we had got some more letters of Jeffrey's, always sprightly, and always wise so far as they go.

The permanent charm of the book is in its picture, or varying pictures, of the career of so much industry, sense, and virtue, from unassisted obscurity to splendid moral public power. But is he not too uniformly wise? If you come to a second edition, give us any of his wild letters if they exist, and refresh us by twenty letters from Jeffrey.

Ever,
H. COCKBURN.

———

Letter from Miss Maria Stewart† to Mrs. Horner.

Catrine, 11th April, 1843.

MY DEAR MRS. HORNER.—My own feelings in former times make me well aware how welcome any praise of a book from a really good judge is to an author's *family*, however well satisfied they may themselves be, that the book needs no

° Volume II, p. 322.
† Daughter of Dugald Stewart

praise to prove its merits. And the praise used always to be more welcome to my mother and me when it came to us *indirectly*, not originally intended to reach us. I am tempted by these vivid recollections, to copy for *your* benefit and that of your daughters, part of a letter I received to-day, from Lady Anna Maria Donkin. She is not a person apt to bestow lavish commendation, and I know few people on whose judgment I would rely more in such a case as the present. She says, "Though hardly able to *see* the book, I have been absorbed in the 'Life and Correspondence,' it has been to me a source of *pure* and *unalloyed* delight. It *must* do good, because such absolute proof that goodness, virtue, high mindedness, principle, modesty, and talent, are *possible*, even in this world of baseness and wickedness, must be beneficial to the young. And as there is not one word of argument, not a sentence that is not the natural expression of thought and feeling without any attempt to teach or lay down the law, but all is said under a conviction that the sentiments are those common to all good people, there is nothing to provoke dispute, but on the contrary, everything to lead naturally to truth and virtue. I think Leonard's part perfect. He has perhaps published some unnecessary letters, but he has not obtruded himself or made *one* unnecessary or incongruous remark. Another great pleasure to me in this book, is the resuscitation of Lord Webb. I feel as if I heard his voice again, after an interval of twenty-four years, and he is presented to the world at a time when I thought his name had vanished, and that with a few of us that remain, his memory would vanish too. So those who knew him as I did, and who were more or less under his *moral tuition*, the picture of him and his docile but very superior pupil, is quite delightful. There is, however, one letter of his (the last) which I wish had not been published. The answer is full of magnanimity, but I am sorry that Lord Webb's more confined views have been exposed, even though the other shines the more, by the comparison. Then there is Mr. Wilson, my kind good old man, brought back to me, and made immortal, for depend upon it the letters of Mr. Horner never will die."

I do not add any remarks or any praise of my own, I am too much interested in the whole work, too much gratified by the appreciation of my father's character, and too much accustomed to consider Mr. Horner's merits and fame as things in which *we* were personally interested, to suppose myself capable of forming an unbiassed opinion. I listen with almost as much eagerness as you could do to catch what others say of the book. All I have as yet heard is delightful. All who have mentioned it to me or in my hearing, have made use of *nearly* the same language as Lady Anna Maria Donkin. The same admiration of Mr. Horner's talents and virtues, the same approbation of the good taste with which *your* Mr. Horner has executed the work, and the same conviction that it must produce a good effect on the world, especially on the young.

Give my kindest regards and good wishes to Mr. Horner and all the young ladies.

And believe me always,

Yours affectionately,

M. D. STEWART.

To his Wife.

Lancaster, *September 9th*, 1843.

I sent off a letter to you yesterday from Manchester, but I sit down that I may have the pleasure of this blessed substitute for personal communication, for then I seem to have you more directly with me in thought, though that is very often the case in the course of the day, and at night after thanking God for his numerous blessings, and above all His preservation of my dearest wife and children, I think on them, and fall asleep thinking of them.

The account from the French of our dear little Queen's visit is interesting ; it seems to have been successful in every way, and her reception in France, and on her re-landing in England, must have been very gratifying,|not to herself only, but to those who witnessed it. When the Monarch is so perfectly respectable in conduct as ours is, these demonstrations of attachment are very reasonable, and tend to good order ; they are only reprehensible when they are exhibited to a person undeserving of regard, either from personal bad conduct, or of conduct injurious to the welfare of the subject. I hope, too, that the occurrence will tend to increase a good feeling between the countries, and the *amour propre* of the French will have been gratified by the monarch of England coming to them, and that monarch a Queen.

My book is Humboldt's new work, " Asie Centrale, which Brown has lent me, and I find it very interesting and curious.

Monday, September 11th.—On Saturday evening I walked to the castle, it was a very warm balmy air, and the full moon shone out brightly. I looked upon it and fancied that you might then be doing the same. Yesterday I passed in writing, reading, and walking, or rather strolling. I did not go to church, but I read *three* sermons of Robert Hall, and a part of Cicero's disquisition on the immortality of the soul, in his Tusculan Questions, so I was not without my serious reflections on the day of rest. I had Madame de Stael's " Allemagne," for another part of the day, and when candles came, I took to the larger print of Humboldt ; thus though there was profound silence, I was in good company all day, with very pleasing and profitable converse. This morning was very cloudy and a heavy rain made me fear that the fine weather was going just when I most want it, for I shall be driving about for some days in an open car probably, but it cleared off and the

weather is again fine. I have been a round of about eighteen
miles, through pretty country, and have seen some very clean
and well ordered mills. John Hutton Fisher stopped me in
the road, I shall call upon him at Kirkby Lonsdale to-morrow
evening. I was in a comfortable little fly and I read the two
first books of the " Paradise Lost," by the way, such majesty
of diction is not to be found in any other work in our tongue.
I expect to go by the mail to-morrow morning to Kirkby
Lonsdale, and then proceed in a car to Sedbergh.

To his Wife.

Gawthorpe Hall, 24*th September*, 1843.

On Friday I was out all day visiting mills by the road side
from Haslingdon to Burnley. where they are thickly set, and
got to Burnley by five o'clock. Yesterday I was engaged in
the town all morning, and about four o'clock the Kay Shuttleworths (Mr. and Mrs.) called for me in their open carriage
and brought me here, about two miles from Burnley. I am
very much struck with the place. It is a very perfect and
entire specimen of an old English mansion, large and lofty,
in what is called the Elizabethan style. It is supposed that
the external walls are of an older time than the dates which
are found on the plaster of the ceilings in several places, viz.,
1604. Except the tower staircase, which is a modern structure,
the rest of the house is in its old state, oak panellings, large
fireplaces with raised hearths, the dining-room has a dais
with oak seats all round, where the master of the house sat
with his friends, the tenants a step lower, and at the other
end there is a large and handsome oak music gallery. But I
will not proceed with my description, for it would not give you
a better general idea of the mansion than the specimens I
have already described. I must add, however, that there is a
fine old gallery with family portraits, which put me in mind of

Haddon Hall. One of the pictures is a fine old head, a
Lawrence Shuttleworth, who was Solicitor-General to Crom-
well. The house stands in a park, and the view of the rising
land opposite with Pendle Hill, 1,600 feet high, behind is good.
As we drove from Burnley yesterday, we stopped at a school
recently built for the neighbouring village by my hostess and
her husband.

I found three brothers of Shuttleworth in the house,
pleasing, gentlemanlike young men. The evening passed very
pleasantly, but Shuttleworth was absent for above an hour to
meet a set of men in the village, who are to form a band. He
has brought down a complete set of instruments, of a military
band, and has engaged a master, another source of happiness
and innocent occupation for the people. I have still another
trait of their benevolent wisdom. To encourage the people to
cultivate a taste for gardening, and at the same time give
them pleasure and profit, he has divided a field of four acres
into sixty allotments, which gives one-fifteenth of an acre to
each, and this is let at a rent of four shillings a year each,
payable in fourpences every fortnight, from April till
September, the time of year when most money is coming in,
and they have least occasion for extra expenses. There are
54 allotment-tenants, a few having more than one. They
cultivate all sorts of vegetables for their families. After a
pleasant breakfast, preceded by prayers, Shuttleworth and his
brothers went to the Sunday school, and Mrs. Shuttleworth
and I had a pleasant chat, and she took me over the house,
told me old stories about it, shewed me places that had been
made for concealment behind the panels in the time of the
Civil Wars, and some old carved wood with names of her
ancestors, and dates in the reign of Richard II., they have a
regular genealogical descent from that time. Having thus
been shewn the oldest of the Shuttleworths, I asked her to
shew me the newest, her baby, about four months old—a girl,

—and a fine child it is. When we came home, we found Mr. Hullah, whom they expected. He is much respected and liked by them. He has been making a musical tour in Lancashire, but is returning to London in a day or two. After luncheon we walked up to the school-house, where there was evening service, and a sermon.

I am now come up to my room, after having been listening for some time to Mr. Hullah playing and singing some sacred music.

It is a beautiful clear star-light night.

CHAPTER IV.
1844—1845.

To his Wife.

May 26th, 1844.

I was at Gorhambury when the division on the Factory Bill took place, and Government got the extraordinary and unexpected majority of 135. Nothing could exceed the universal astonishment, and many of their supporters grumbled much at having been compelled to vote with them or stay away, without any necessity. But they were wrong, for it was of great consequence to get such a majority as should put an end to the question, which this has done. It was a great triumph to Graham, who deserves it; for his conduct on this occasion, and the ferocious personality with which he has been assailed, have conciliated the sympathies of many even of his enemies and opponents. But no man ever rose so much as he has latterly done. His capacity and administrative powers are admitted by all to be first-rate, and he has evinced so much more of temper and moderation, avoiding giving offence by a bitter and sarcastic tone, that he has disarmed a great deal of hostility and aversion, and there is a general disposition to do justice to his firmness, ability and honesty on this occasion. This division is also a pregnant proof of the strength and power of the Government, when they choose to exert it; and when their position now is compared with that in which they were placed at the end of last session, everybody must see how enormously they have gained.

*To his Daughter Frances.**

Bedford Place, *22nd May,* 1844

MY DEAREST FRANCES,—In compliance with your request, I

*Just before her marriage to Mr. Bunbury.

am now going to give you a discourse on family accounts, a
dry subject certainly, but one in which our comforts and
respectability are materially involved.

No prudent person, however great his fortune, should
neglect to keep an account of the manner in which his annual
income is expended; to persons with moderate means there is
no safety without it, and to those who are obliged to practise
a strict economy, it is the greatest safeguard, for it is like a
confessional, whereby many thoughtless expenses, which swell
to a large amount at the end of the year, may be checked. I
need say little to you of the importance of avoiding debt; I
mean tradesmen's bills, for you know that the practice in your
present home has always been to avoid them as much as
possible. You cannot altogether avoid having a bill with
some tradespeople, but there is no reason why it should ever
exceed three months. Always have a list of the names of all
the tradespeople you deal with, where you have accounts
standing against you.

At the beginning of every year let it be an indispensable
rule with you to make an estimate of the way in which you
propose to spend your income. It can only be a rough one
until you have had the benefit of experience, but a very rough
one is better than none at all. After you have set down such
sums as you cannot escape from, next take those of most
necessity, and having ascertained what these will amount to,
you will be able to judge what you have to spend upon
indulgences.

It will be always prudent to lay aside a full sum for medical
attendance; if it has not been spent, it forms a very nice fund
on which you can draw for indulgences in the following year.
In keeping accounts a rigid accuracy is indispensable. The
minuteness with which particulars of expenditure are to be
set down, will depend upon the amount of your means.
Persons in easy circumstances may take sums which they

may set down in the gross, under the head of "pocket money," and keep no account how it goes; but for those to whom a rigid economy is necessary, who require a check upon trifling outgoings, a setting down of minute sums is quite indispensable.

In what I am now saying I refer to *family accounts only*. Two books only are necessary, a cash book, and a ledger, and these I have provided for you.

In your cash book put down whatever you receive, never mind from what source, put it down. In like manner, enter whatever you spend; it is easy to separate the items afterwards in whatever way is necessary. Balance your cash book at short intervals, never longer than a week. As there may be some days on which you may not spend anything, it will render the keeping of your cash book more easy if you in such cases add against the date the word "nil."

You should have a fixed time for your accounts, for regularity's sake, immediately after breakfast is best.

When you are from home and cannot have, conveniently, daily access to your cash book, note down your expenses in a pocket memorandum book.

I have divided your ledger into the main heads of expense in every family, and once a month you should transfer the items from your cash book under the several accounts in the ledger.

Some of the above directions may appear trivial, but accuracy and a satisfactory review at the end of the year depend upon them. I will conclude with Crabbe's excellent definition of a wise economy :—

> "The wise economy that does not give
> A life to sparing, but that spares to live.
> Sparing, not pinching, mindful but not mean,
> O'er all presiding, but in nothing seen."

I am, my dearest Frances,

Your affectionate father,

LEONARD HORNER.

To his Daughter.

Darlington, 18*th June*, 1844.

MY DEAREST LEONORA,—This day the railroad that unites
Darlington and Newcastle was opened, and there is a great
meeting at the latter place in celebration of it. A special train
left London this morning, and I read in the station at York
at half-past eleven to-day, the *Morning Herald* published in
London this morning, giving a full report of speeches that
were delivered in the House of Commons eleven hours before.
This is one of the greatest feats of railway travelling·
Newcastle is now only thirteen hours from London of ordinary
railway travelling. I remember when it was forty-eight hours
by the mail.

I am happy to hear that there was so fine a day for Chiswick,
and that Babbage's party was so pleasant. I hope he was
invited to meet the King of Saxony at Sir Robert Peel's
among the other scientific men. I am so glad to see that the
modest Mr. Brown was there. I regret very much that it
was impossible for Charles to go. It is very creditable to Sir
Robert Peel to take that marked notice, as Prime Minister, of
scientific men. It is something new in this country.

I regret that I did not hear the last lecture of Faraday, and
only wish I could follow a long continuous course of chemistry,
under so able a lecturer.

My kindest love to your dear Mamma and sisters.

Your affectionate father,

LEONARD HORNER.

To his Wife.

Edinburgh, 24*th June*, 1844.

We had a pleasant sociable dinner at Bonaly, Cockburn
merry and playful as ever. It was a beautiful day and very
warm, and at eleven o'clock at night he and I walked on the

terrace, where it was so light that I could easily have read; the light was so strong that very few stars could be seen, and the air quite balmy. Little did I expect so sudden a change. To-day there is a cold east wind and a heavy rain, which renders my moving about impossible. The rain, however, came on only about two. Cockburn and I started in the drosky at eight, and coming round by Edinburgh got to Craigcrook at half-past nine. We had a pleasant breakfast. I have since that been calling on Thomas Thomson, whom I saw.

I have been interrupted just now, by the arrival of four gentlemen, a deputation from the Directors of the School of Arts, to request me to accept of a public dinner in testimony of their respect for me as the founder of the School of Arts, and with a message from the Lord Provost that he will be gratified in taking the Chair on the occasion.

This is certainly a very flattering and unexpected honour, and one which I could not decline. I have named Friday se'night as the day most convenient to me, after I have been to Kinnordy.

I shall be disappointed if I do not see Sir George Napier before they leave England. I am grieved to hear of the sad relapse of Miss Fox, and fear it must be all over by this time. We shall often think, with pleasure of our evening visits to her, this time last year.

———

To his Daughter.

Edinburgh, *27th June*, 1844.

MY DEAREST KATHARINE,—I had hoped to write a long letter before breakfast, but Lord Cockburn and Lizzy were ready before nine, and that done, in comes Lord Fullerton; when he had gone, it was time for me to go, according to an appointment, to call on Hugh Miller (the old Red Sandstone), who had so many interesting fossils to show me that I could not get soon away. He is a very remarkable man.

Tuesday.—I found the Academy in a very flourishing state. Perfect in order and everything with the air of business and success. I heard the Rector examine his boys in Virgil for an hour, and I do not think it is possible there can be better teaching. They are a remarkably gentleman-like set of boys. He announced me to them as one of the founders of the Academy, and they gave me a round of applause.

We had that day a very nice party at the Maitland's, but I was obliged to go away early to sup with Mr. Bryson, the watch maker, to meet some of the Directors of the School of Arts, very sensible men.

Nothing can exceed the kindness of everybody to me, and I am enjoying my visit to my native town, and to my old friends, very much.

God bless you all,

Your affectionate father,

LEONARD HORNER.

From Lord Jeffrey.

Craig Crook, *Friday Night, July 5th,* 1844.

MY DEAR HORNER,—I hope you understand that nothing but the shabby state of my health could have prevented me from witnessing and enjoying your well earned triumph of this evening, for which if I do not envy you, it is only because I feel that it is not a gift of fortune which might as well have come my way as yours, but the just reward of services in which I had no share. I do, however, rejoice in it almost as much as if both the merit and the reward had been my own, and I must add, that, next to my cordial sympathy in the gratification it must have afforded to you, is my satisfaction in seeing that you are still remembered as you ought to be

by those to whom you have been so great a benefactor; and
that neither the lapse of years, nor the political and religious
feuds of which they have been so prolific, have deadened the
sense of the obligation which the middling and working
classes of our population must always continue to owe you.
But you are sick, I daresay, by this time, of such felicitations,
and I have not chosen my time well, I fear, in following up the
thunders which are probably rolling around you at this
moment, with the " still small voice " of my humble sincerity!
My apology, however, must be, that I really could not help it,
and that I write much more for my own gratification than
for yours.

If you have come safe and sound out of this great trial, it
would be a great pleasure to us to see how you look after it?

But in the meantime, and at all events.

Believe me always, my dear Horner,

Very faithfully yours,

F. JEFFREY.

From Lord Cockburn to Mrs. C. Lyell.

July 6th, 1844.

MY DEAREST MRS. LYELL,—I have but a moment. . . A
capital effusion last night. Perfect. Above a hundred present.
All superior, model people. Had the room been large enough,
triple the number could have squeezed in. Horner's address,
in response to his own health, was as good as *possible.* Honest,
simple, judicious, affectionate, and if wrong in point of time,
only too short; a rare fault, into even which however, he did
not fall. Rutherford's memory of Francis Horner was
absolutely perfect. Nothing is so difficult to do, and the
memory of no man revered for mere virtue, judgment, and
bright prospects prematurely closed, and with no published
works or historical achievements to descant upon, was ever
given in better taste, or with more exquisite thought or diction.

There was only one cloud, or rather wet blanket; consisting of a sermon, *at the least an hour long*, by Prof. Wilson, on poetry, rent, Scotch peasantry, Malthus, the French Revolution, Ricardo, tile draining, moral philosophy, cottage windows, penny postage, and about a dozen of equally connected topics. But the sun woke out again, and on the whole the meeting was excellent; the prevailing feeling of respect and affection for your father was most delightful, and came like cream over my heart. No private citizen could have received a more gratifying homage, from all parties, and for truly good works. I thought of you all, with interest and love. Remember me to Lyell and everybody. Frances is to be with us, I understand, on Sunday s'enight, when I shall be glad to introduce Mr. Bunbury to the Pentlands.

God bless you.

Ever,

H. COCKBURN.

To his Sister.

Mildenhall, *September* 15*th*, 1844.

MY DEAREST NANCY,—I am very happy to have an opportunity of writing to you from the house of my dear Frances. We left London, Anne, Katharine and I, on the 11th, travelled by railway to Bishop Stortford, and by coach from thence to Bury, and found Lady Bunbury waiting in her carriage for us, and in twenty minutes we were at Barton Hall. Charles Lyell and Mary went to Mildenhall on the 7th, and came to Barton with Charles Bunbury and Frances on the 11th to meet us. We remained till yesterday, and Mamma, Kate and I came here with Frances and her husband.

It has been a very great pleasure and satisfaction to me to spend these three days at Barton. I never met with persons in whom I found more to admire, respect, and love, than Sir

Henry and Lady Bunbury. They are both persons of excellent
understandings by nature, and extensively cultivated; all
their sentiments are characterized by good sense, liberality,
moderation, and benevolence. He has the reputation of being
one of the best landlords in the county, and from what I saw
and heard, I can well believe it. For more than twenty-five
years he has followed the allotment system with his labourers,
and increased the number of allotments year by year, as he had
more and more experience of the good effects of the system
upon his people, physically and morally. He builds excellent
cottages, and to each there is a garden, and half an acre of
land. I visited several of these, went through the houses,
and talked with the people. The cottage has two stories ; on
the ground floor, a living room and a kitchen, in the latter
there is an oven and a copper ; above are two well-aired, well
lighted bedrooms, all the rooms are plastered and whitewashed.
I came quite unexpectedly, and there was not a room that was
not scrupulously clean, and in order, and sweet smelling, and
the furniture good and in order. They have each a little
court with a place for a pig, which they all keep, and for
rubbish, &c. Their garden is about one-eighth of an acre,
and they grow vegetables and flowers, and their half acre of
land is occupied with wheat, turnips, pease and potatoes.
They pay for the whole from £3 3s. to £3 10s. a year, which
rent is collected once a year, at Michaelmas, when, from the
extra pay in harvest time, the most money is coming in, and
Lady Bunbury told me that there never has been an instance
of any one not being ready with his rent on the day. Sir
Henry has around Barton Hall 120 of these cottages and
allotments, and around Mildenhall more than 200. I should
tell you that he engages a well educated medical man in Bury
to attend on these cottagers and their families, for which they
pay 5s. a year (included in the above £3 3s. or £3 10s.) and
he adds 1s. more for each, and for this they have medicine

and attendance, except in cases of childbirth, when they pay
extra. There is a school at Barton, and another here, maintained by Sir Henry. Besides these great moral and
benevolent qualities, which his care of his people bespeaks, he
is a man of great accomplishment and refinement, and has a
very noble library, with which he appears to be well acquainted.
There are many fine pictures in the house, and among them
twelve Sir Joshua's. There is a full length picture of Lady
Sarah Napier when she was young, which Sir Joshua said,
when he finished it, was the finest he had ever painted. There
is an exquisite one of Sir Henry's own mother. Sir Joshua
was Sir Henry's godfather, and he told me that he remembered,
when a boy at Westminster School, often visiting Sir Joshua,
he remembers seeing him painting the Infant Hercules, and
the living snakes in the room which are introduced in the
picture. The house is very large and well furnished, but
without display; it stands in a fine park, with many noble
trees, two oaks on the lawn are splendid, one is seventeen and
a half, and the other is nineteen feet in girth. There are
large fruit and kitchen gardens and an arboretum. Every
Sunday evening his grounds and gardens and shrubberies
close up to the house are open to his labourers and their
families to walk in, which they enjoy greatly.

Of Lady Bunbury I could speak with pleasure and praise
for a long time. I saw much of her, and every day I saw
more to admire in her, she is worthy of being the daughter of
Lady Sarah Napier, the niece of Lady Louisa Conolly, and
the sister of Generals Charles, George, and William Napier,
and higher praise I could not give her. The sons, Edward,
nominally a barrister, and Captain Henry, whose regiment is
in New Brunswick, and from which he lately came, were at
Barton, and most agreeable men they are; and Cecilia, Sir
George Napier's daughter, almost adopted as their own by Sir
Henry and Lady Bunbury, is a very engaging girl. Nothing

can be more kind and affectionate than every member of the
family is to Frances, and I see that she is on the most easy
and happy terms with them.

I must now say something of Mildenhall and its inhabitants.
I am greatly pleased with the house, it is old-fashioned, but
the rooms are excellent and most comfortable, and Sir Henry
has had it fitted up most nicely for them. There is a large
lawn of shaven turf, a good deal of wood round the house, and
a paddock large enough to feed two cows. There are some
large trees in the paddock that must be above a hundred
years old, and there is a very fine walnut tree in the kitchen
garden. We had excellent grapes, peaches, plums, and pears,
at dessert yesterday, from their own garden. Dearest *Fanny*,
for that is the name she goes by, has everything around her
to make her happy, and she is perfectly so I believe—an
admirable, sensible and accomplished husband, who dotes on
her. I am sitting in a large room hung with tapestry, with
the original picture of Catharine of Braganza, sent over to
Charles II. before he married her. Sir Thomas Hanner, one
of Sir Henry's ancestors, was Cup-bearer to Charles I., so
that the picture they have, came into the possession of the
family in that way.

We intend to remain here till Thursday next, when we go
to pass a few days with the Nightingales in Derbyshire, on our
way to York.

Adieu, dearest Nancy, with best love from all here.

Your affectionate brother,

LEONARD HORNER.

To Charles J. F. Bunbury, Esq.

London, 1st *December*, 1844.

MY DEAR CHARLES,—I have been happy to hear that fossil
botany interests you so greatly, not only on account of the

satisfaction it must give to yourself to find that you see before you a field of great promise for the advancement of science, but also that besides the immediate pleasure which the investigation of the subject affords, you have that of working to a good purpose as a geologist. I am confident that, without taking an undue share of time from the other important duties which belong to your position, you will do much for science ; and I have no doubt that we shall see you ere long, one of the recognized authorities in fossil botany, to whom geologists are to look up in that department.

You have heard that my friends at the Geological Society have conferred the great honour upon me of thinking me worthy of occupying the chair at the next election. I yesterday had a very satisfactory conversation with my superior officer, Sir James Graham, on the subject, who said many kind things, and gave his entire consent to my accepting the office.

I should add that before I did consent to take the office, I had quite satisfied myself that it need in no respect interfere with these duties.

All my leisure therefore, from henceforth to the end of the term of the presidency, must be almost exclusively devoted to those things that are most likely to make me a useful President to the Society, and there are many ways in which I see opportunities of usefulness. It is now thirty years since I acted as secretary, and it is a great privilege and a great delight to me, to be able after so long an interval, to resume official activity in the service of the Geological Society. I feel the *veteris vestigia flammæ* very distinctly within me.

Fanny must be doing much good by the time she is giving to the school, but I am glad to learn that she is laying aside a portion only of her time for that and other objects of the kind. A moderate part of the day given regularly, will, at the end of the year, amount to a large sum ; quite as much as any sense of duty can call for. It would be a great pity were she to lose sight in any degree, of other objects she has hitherto been much occupied with.

I have read some of Henslow's late communications to the Bury newspaper, which he has been so good as to send me. While I agree in much that he says, and cannot too greatly admire his devotion to so good and important a cause, I cannot assent to the principles he has laid down, that in order to maintain the labouring classes in the agricultural districts in a proper state, a certain amount of labourers must be assigned to a certain amount of land, and that the labourer must be first considered, and then the farmer. I consider that principle to be quite unsound. A man takes a farm, that is, agrees to pay a yearly sum to the owner of a piece of land, with no other object in view than to make money by it, and as much money as he can, taking an enlarged view of what is his interest in the long run. He is then in the same position as every other man who follows a trade or profession, to procure his subsistence, comforts, and luxuries. There is no more obligation on the farmer than on any other member of the community of equal means, to consider the state of his less wealthy fellow men around him. He is not only perfectly entitled to take all fair means to produce the commodity he has to sell as cheaply as possible, in order to make his profits as large as possible, but he is compelled to do so by competition, for otherwise he could not stand his ground. No one man can regulate the rate of wages, and no possible combination of men can do it, so long as competition is free. If there is a redundant population in a low moral state, and free competition, there will be a constant tendency to the lowest rate of payment that will sustain animal life, and there is but one cure for the dreadful evil of a country swarming with mere human existences in my opinion, viz., bringing out the higher qualities of these existences; in other words, raising the moral condition of the population to that state of right

feeling that people will not marry without the knowledge that
they can keep up themselves and their offspring, in the exercise
and enjoyment of those powers by which man is raised above
mere animal existence. To bring about that higher moral
state of the great mass of the population, there must be an
extensive call upon that portion of the community where there
is the greater amount of wealth, in order to provide those
things which no effort of the less wealthy ever can provide : good
dwellings, good schools, good religious instruction, good amuse-
ments, good government. But on the other hand, when
things have got into so bad a state as they have been suffered
to fall into, temporary and extraordinary means must be
resorted to, and therefore I might be disposed to go along with
Henslow, and say that the first thing to be done is to provide
employment for the people. But my paper tells me it is time
to stop, and you will perhaps say, " I wish it had told you so
long ago."   So with love to my dear Frances, I conclude.

I am, my dear Charles,

Affectionately yours,

LEONARD HORNER.

———

To Charles J. F. Bunbury.

London, *June 8th*, 1845.

MY DEAR CHARLES,—Your letter gave me much pleasure,
as you appear to have liked your visit to Henslow so much,
and to have profited so greatly by his skill as a fossil botanist.
You have a fair field open to you, in which great good to science
is to be gained and honours to yourself won, and I have no
doubt to see both fruits ripened abundantly by your culture.
I long to hear all particulars from your own lips, and to take
a lesson from you.   We had an announcement at the Society
on Wednesday of a *Sigillaria* having been seen at St. Helens
in Lancashire, with roots passing into Stigmaria.   The last,

meeting of the season will be on Wednesday, when there will
be an excellent paper on the geology of the Isle of Man by Mr.
Cumming. He thinks he has discovered evidence of trap
eruptions as recent as the period when the boulder clay was
accumulated. I shall be very curious to see his evidence
which is to be contained in a future paper.

I look forward with great pleasure to next Saturday, when
we hope to see our dear children at Mildenhall. The peace-
fulness of the three following days will be delicious, and we
shall have together a very agreeable week at Cambridge* I do
not doubt.

I strongly recommend your father to go, and as he did not
seem disinclined, I have had rooms secured for him in Trinity,
in the same staircase with Edward. I have written to him to
that effect this day; I think that he would be interested, and
see many friends.

Love to dearest Fanny,

Your affectionate father-in-law,

LEONARD HORNER.

To his Daughter.

Church Stretton, *7th August,* 1845.

MY DEAREST KATHARINE.—Yesterday morning I left Ludlow
about nine, and arrived here between three and four. This is
a small market town, in a narrow valley, with the lofty range of
the Longmynds on the west, and a lower but very picturesque
range of detached hills, called Caer Caradoc, on the east. The
latter are of igneous origin very similar to those surrounding
you, Arthur's Seat, Costorphine, and the Pentlands. But while
these last are mantled round by the coal-bearing beds,
Caradoc has Silures around him, a more ancient race than
the sooty gentlemen that attend upon Arthur. With the

* At the British Association.

exception of a few showers, this has been a very fine day, and
I made a round of the Caradoc Hills. I could find no other
conveyance than a chaise, and as the round was at least
twelve miles, a donkey could not have accomplished it in less
than two days probably. To-morrow I mean to bestride one
in ascending Caradoc, which is, I think, about one thousand
two hundred feet.

My dearest Katharine's affectionate father,

LEONARD HORNER.

———

To his Wife.

Bridstow near Ross, *August* 10*th*, 1845.

I wrote last to Katharine from Church Stretton. You have
heard me speak of the Rev. Thomas Lewis, as an eminent
geologist of this part of England, and I wrote to him from
Ludlow asking him for some information as to places I ought
to go to. At Stretton I received from him a kind letter, and
two hours afterwards, a messenger brought me a very obliging
note from Mrs. Acton. I wrote to Mr. Lewis that I would
breakfast with him here this morning, and I was reluctantly
obliged to decline the invitation of Mrs. Acton. Friday was a
very fine day and I got to the top of the Caer Caradoc, from which
there is a most splendid view. I saw the Wrekin at no great
distance from me, and regretted that I could not be exploring its
sides. At five o'clock the mail came up to Stretton, I found
a place, and came on to Hereford by Ludlow and Leominster,
forty miles, getting in about nine o'clock. Yesterday morning,
by the merest accident, I met Mr. Lewis in the yard of the
Inn, who had just arrived by a coach from Amystry and was
going immediately to this place. He left the coach and came
by a fly with me, to my great advantage, for between Hereford
and Ross there were several points of great interest, which he
from his intimate acquaintance with the locality was able to

direct me to, and we were seven hours going the thirteen miles, so you may conclude that there was much to be seen. Independently of its geological attractions, it is a most beautiful drive from Hereford to Ross, particularly in the neighbourhood of Home Lacy, where the old Duke of Norfolk (of our time) used to reside. From the heights above the grounds, with a fine reach of the Wye flowing through emerald meadows, there is a vast prospect bounded by the Welsh Mountains. I have written to Hallam, offering to go to him on Thursday. If he can receive me, I shall stay with him till after breakfast on Saturday, as I mean to return home that day.

To his Daughter.

Liverpool, *August* 31*st*, 1845.

MY DEAREST KATHARINE,—Having been reading a good deal of Blanco White (with which I am more and more pleased and edified) and remembering his account of his going to Paradise Street Chapel, and hearing Mr. Martineau preach, I went there this morning, and had the good fortune to hear him (the first time after an absence of several weeks that he had preached), and a most admirable sermon he gave, one which a powerful mind only could have composed, setting forth the duty of a life of active exertion and combating with difficulties, and contrasting the superior pleasures which a retrospect of a life so spent affords, above what can possibly be left by a life of indolence and present ease.

I am curious to see the editor of Blanco White's book, Mr. Thom, and mean to go to his chapel this afternoon. I shall be here till Thursday, having a good deal of factory work to look after here.

I should have liked to have seen the Danseuses Viennoises, for it is to me as interesting a thing to see young creatures dancing, as it is the reverse to see the posture making of men and women in a ballet.

I hope you have made good arrangements for your getting safely to Strachur.

This is a beautiful day, and I trust there is now a good chance of a good harvest.

To his Daughter.

Liverpool, *4th September,* 1845.

MY DEAREST SUSAN,—My letter to Joanna announced the arrival of Charles and Mary yesterday.   I had a happy evening with the dear creatures. We breakfasted at eight, and at ten we left the hotel for the "Egremont ship," found several passengers there, among them Mr. Everett.   All was put on board a small steamer, and I went along with them to the " Britannia " two miles off. I went with Mary to her cabin, and stayed about half-an-hour on board, and then the little steamer hauled off, and I waved my hand to dearest Mary, standing with Mrs. and Miss Everett on the high deck. Everything at present looks favourable for their voyage.   God speed them.   It is possible they may reach Halifax in time to send a letter by the steamer that leaves Boston on the 15th. They are both uncommonly well, and go with good hopes and spirits.

My love to all,

Your affectionate father,

LEONARD HORNER.

To Charles J. F. Bunbury, Esq.

Manchester, 18*th October,* 1845.

MY DEAR CHARLES,—I.was much pleased to learn that you have prepared a paper for the Geological Society, and if you will send me the title, I will have it entered, that it may be read soon. I have asked three or four persons who live in the neighbourhood of coal mines here, to send me any fossil plants that are found, and they have promised to do so. They are destined for you in the first instance, and for the Society after you have taken what you want for your own collection.

You will read to-morrow a delightful letter from Mary. How instantly they got in the midst of friends and of business. They certainly do not let the grass grow under their feet.

I was much shocked to-day by reading the death of Mr. Basevi, the architect, by falling from the tower of Ely Cathedral, where he was standing with the Dean. What a terrible shock it must have been to the Dean. I knew Mr. Basevi, he was an amiable and excellent man, and of very high merits in his profession.

I am, my dear Charles,

Affectionately yours,

LEONARD HORNER.

To his Daughter, Mrs. Lyell.

Bedford Place, *Sunday Evening, November* 30*th,* 1845.

Here I am in my study my dearest Mary, just come down after reading one of Orville Dewey's beautiful, eloquent discourses, that entitled, " Life is what we make it," which contains a few pages more sound philosophy and more practical wisdom, than is to be found in many volumes of sermons. When I came home yesterday to dinner, I was met by the joyful voice of Nora—" A letter from Mary ! " Every letter you and Charles send, makes me feel New England more and more as a part of Old England, and increases my desire to see it, a vain desire, which can never be gratified. Were it not for other uncontrollable circumstances, the waves of the Atlantic would not deter me.

Charles Bunbury has written an excellent paper on the coal plants of Frostburg. The number of the *Westminster Review*

just out, contains an article on Blanco White's Biography.
Mr. Thom told me that it was coming, and that it is by Baden
Powell. I have not had time to do more than cast my eye
over it, but it appears to be in a kind and fair spirit, and to
give due praise to Mr. Thom's fidelity as executor and editor.
I talked with Baden Powell at the Royal Society last week
about Blanco White, and he spoke of him with the greatest
regard.

My dear Charles,—your paper on Auvergne was read on
the 19th and discussed by Hamilton, Sedgwick, Forbes and
Mantell. Your letter of the 1st came in time for our last
meeting, I read all the geological part at the club, to the
great amusement of all present in what you say of Koch's
Leviathan, and at the meeting I read what you say about
the perfect skeleton of the Mastodon. Owen was not present,
but Mantell got up and told us of the letter he had received
from Silliman, describing a very remarkable fossil animal
recently discovered by M. Koch, quite gravely, upon which
Hamilton and some of the Clubists said in a loud whisper
" Pray read the rest of Lyell's letter," so I got up and
said that Mr. Lyell had alluded to the remarkable fossil beast
of which Dr. Mantell had spoken, but as his communication
was not exactly of a kind to read before a grave Society, I
should dissolve this meeting and consider ourselves a social
party, and amuse them with Mr. Lyell's account of Mr.
Koch's discovery. I then read what you say, and much to the
entertainment of those present, Mantell joining heartily in
the laugh. I called on Owen, and gave him the extract of the
geological part of your letter, and it has produced a communication from him to you on the Mastodon, which shall go
by the same ship as this.

Next Wednesday, besides Bunbury's paper, we are to have
a description by Mantell of some bones of the Iguanodon dug
out of the Wealden beds, in the Isle of Wight, within the last

three weeks, one of them a tibia of so enormous a size as to make the brute thirteen feet high, whereas the largest found before, only made him nine.

I am reading Murchison's "Russia," there is a great deal that is very interesting. My chief difficulty, as President, is to find time to read for my anniversary address. Nothing now has been heard of Agassiz ; there is no account yet of his having got to Paris. I think I told you in my last, that Joanna put into my hands a note from Nelson to you, saying that his sister had written to him of a recent fresh sub-sidence in the Basin of Cutch. I wrote to him asking him to send it, he has done so, and here it is.

"One of Captain MacMurdo's guides was travelling on foot to him from Booj last June. The day he reached Luckput there were some shocks which shook down some of the walls of the fort, and some lives were lost ; the sea at the same time rolling up the Kori mouth of the Indus, and overflowing the country as far westward as the Goongra River, northward as far as a little north of Dera, and eastward to the Sindree Lake. The guide was detained six days at Luckput, during which sixty-six shocks were counted : he then got across to Kotree, of which only a few small buildings on a bit of rising ground are standing. Most of the habitations through the district must have been swept away, considering that the best houses in Scinde are generally built of *sun-dried* bricks, and that whole villages are only huts made of a few crooked poles and reed mats. The guide travelled ten coss (twenty miles) through water on a camel, the water up to the beast's body ; of Lak, nothing was above water but a Fakeer's pole (the flagstaff always erected by the tomb of some holy man), and of Veyre and other villages, only the remains of a few houses were to be seen. Captain MacMurdo tells me that there are generally two earthquakes every year at Luckput. The Sindree Lake has of late years been a salt marsh. We had one remarkably

high tide here, which overflowed part of the road between the
camp and the bunder (quay) ; as nearly as I can now make
out it was at this same period of very stormy weather that the
earthquake occurred."

God bless you both,
Your affectionate father,
LEONARD HORNER.

CHAPTER V.

1846.

To Charles Lyell, Esq.

Bedford Place, *January 26th,* 1846.

MY DEAR CHARLES,—I suppose you must be by this time at New Orleans ; but you would probably find a good deal to interest you in the wide extent of country between the Atlantic and the Mississippi. You gave us great comfort by the impression left upon you at Washington, that so dreadful a calamity to both countries as war is not likely to happen, however General Cass and the democrats of the Western States may declaim. In our Parliament just opened, there have been the most emphatic expressions from all sides of both Houses, of an earnest desire to preserve peace, and Mr. Pakenham's refusing to forward the proposal for a renewed negotiation respecting the Oregon territory, was condemned by Peel. I do not believe that there is any class of persons in this country who would not lament a war with the United States. I will not attempt to give even a sketch of the events of surpassing interest that have been occupying people's minds here for the last six weeks, on the questions of corn laws and free trade. You will have had, I hope, opportunities of seeing English newspapers and learned that Peel gives up the Corn Laws ; and to-morrow we are to learn in what way, and what besides he is to do. There is, for a time at least, a complete break up of his party, and the Duke of Richmond and the Protection agriculturists are furious. What the next month will bring forth no one can conjecture. Matters will be rendered more difficult if the fears of those best acquainted with

Ireland are realized, for the danger of famine in many parts
of the island, from the failure of the potatoes, is imminent.
March and April it is feared will exhibit a frightful state, not
only for the time, but for the future, with reference to the
seed potatoes for next crop.

I must however, before I occupy more of my paper, tell
you what I have to say about the Geological Society. I have
this evening finished my second perusal of your American
travels, with reference to the notice I have to take of them in
my anniversary address. That address has proved the most
onerous part of my duty. I am naturally desirous of doing
my best, but to do anything satisfactory to my own mind, it
has been necessary for me to employ every hour that I could
spare. The sacrifice I have made has been society, and to a
great degree all other reading. Were it not that it is always
painful to work against time, I do not regret any sacrifice I
have made, for I have learned a great deal, and the deeper I
go into the subjects, the more intense does my interest become.
I never felt my Geological ardour greater than it is at present,
although I am full forty-two years older than when I began
the study. I have read carefully the whole of Murchison's
first volume, 600 close 4to pages, and with the greatest
interest; although it does not open up so new a field as his
"Silurian System," it does him no less credit. It will form a
text book for a considerable part of my address. The second
volume is almost wholly Verneuil's; Forbes says it is very
good, and he is to help me to what I am to say about it.
Owen's last number of his "Fossil Mammalia" will be out in a
week, and will contain his introductory discourse; he will give
me some aid too, so that by dint of work and by taking help,
I hope to get through the task without discredit. I keep
steadily to my duties in Somerset House, rarely a day passing
without my being there from three to five, and so things get
on. The last measure I have got carried has been the

Museum ; to make it really serviceable the first thing to be
done is to make our British collection as complete as possible—
stratigraphically arranged—the fossils of each formation
arranged in the natural order, a full catalogue for each formation, superfluous *rock* specimens to be thrown out; large
Mammalian and Saurian remains to be sent to the British
Museum.　When we have got the British collection in order,
we shall then consider what is to be done with the Foreign,
but it will be half a year at least before the British is put in
order.　I have got John Carrick Moore to be our secretary,
with Hamilton, to the entire satisfaction of the latter, and of
the whole Council.

I told him he must be at the Society's house two or three
times a week regularly, and he said that he will cheerfully
give that time at the least.　We have awarded the Wollaston
medal and the balance of the proceeds of the donation fund
to *Lonsdale*, for his "Corals of North America and Russia" as
well as his former works on corals.

At the two last meetings we had a good paper by Sedgwick
on the slate rocks of Westmoreland, greater part Upper
Silurian.　He announced his belief, from recent discoveries of
fossils, that the greater part of the *Llandeilo flags* of South
Wales would be found to be *Upper* Silurian.　We had at last
meeting a paper by Dawson of Picton, on the coal plants of
that neighbourhood, with an excellent supplement by Charles
Bunbury.

February 1st.—We have had great anxiety on account of
Edward Forbes, who has been very seriously ill.　I sat half-
an-hour with him to-day, which is the third time I have seen
him, and he is better but still weak.　I trust, however, that
we have now no reason to be uneasy.

Peel's proposal is to apply a duty to wheat from 10s. to 4s.
as the price varies from 53s. to 48s. and for three years, and
then all duties on the import of corn to cease.　Maize to be

admitted immediately duty free; all duties on meat, cattle,
vegetables, and fruit to cease immediately, and most other
duties to be reduced 50 per cent. Protection amounting to
prohibition on any article to be given up. No new tax, confidence that increased consumption will keep up the revenue.
The agriculturists in a state of frenzy. I mean the Dukes of
Richmond and Buckingham, *et id genus omne*. The friends of
free trade quite willing to accept the terms, but Cobden and
the League still urging an immediate total repeal. The
question comes on again in another week, the best informed
expect that the plan will be carried in the Commons by a
majority of 100, and that the Lords will not in that case dare
to do otherwise than pass the Bill. I am on the whole well,
but too much worked, and I cannot do less. No trouble in
Factory matters, all going on smoothly. Trade very good,
wages good, people in full employment.

God bless you, my dear Charles and Mary. Immediately after
the anniversary I go into Lancashire, to be away five weeks,
in town three weeks, and then to Lancashire again, for a
month, and be in town from 24th May to 18th June, during
which last time I hope to see you return. Such are my
plans, *Deo volente*.

Your affectionate father,
LEONARD HORNER.

To his Daughter, Mrs. C. Lyell.

Bedford Place, *March 1st*, 1846.

MY DEAREST MARY,—The last account we have of you is
that contained in your letters from Savannah of the 9th of
January, when we had the happiness of receiving a most
excellent account of you both. Your account of your visit
to Mr. Hamilton Cooper and his family is very interesting.

You have seen slavery under its most mitigated forms, and

I shall be curious to hear how you find the slave population in the South Western States. We entertain no apprehensions of war on this side of the water, because no one seems to think there can be any considerable portion of the United States population insane enough to wish for that which must bring loss and mischief and misery without any one compensation approaching to an equivalent. Lord Lansdowne told me a few days ago that he was much pleased with the tone of conversation of the American Minister at Berlin, who was lately in London for a short time.

Last Friday, after twelve nights of debate, a division took place on the Government's proposal for the total repeal of the Corn Laws at the end of three years, when they had a majority of ninety-seven. What fight the protectionist party will make hereafter, it is hard to say. If they persist in throwing obstacles in the way of a speedy settlement, they will do the aristocracy and landed gentry great harm, for the protectionists consist almost exclusively of them. There is great reason to apprehend much suffering in Ireland, so that every day lost is a serious matter. There will be a stormy Session, for the high Tories are infuriated against Peel, and will do all they can to vex him. A large number of them voted for De Lacy Evans lately, at Westminster, who threw out Captain Rous, who accepted the place of a Junior Lord of the Admiralty. I hope the Whigs will go to the utmost verge in support of Peel, but there will be a question on which they cannot go with him, the continuance of the higher duty on foreign sugars, and he may be left in a minority on that question.

To-morrow morning I set out for Manchester and expect to be away five weeks. God bless you, my dearest Mary, I must now have a little chat with your husband on Geological matters.

My dear Charles, I spared no labour or pains in the preparation of my address, and the Council have determined

that it is to be printed in the May number of the journal. If
you would like to give away any copies, it cannot be otherwise
than agreeable to me, and I will send you a few for that
purpose. John Moore has entered upon his duties *con spirito*,
he will be an immense comfort to me during my Presidency,
and if he continues for some time, will be invaluable to the
Society.

Your announcement of the discovery of the Megatheriums
at Savannah has excited great interest, as a tooth was all you
had made known before. We had a very good paper last
Wednesday, by Prestwich, on the Tertiaries of the Isle of
Wight. We are to have, shortly, a paper by Darwin on the
Falkland Islands, and Murchison on Erratic Blocks. The said
Sir Roderick was at our anniversary in his star. The
anniversary dinner was well attended, and it was said to have
gone off well.

Milman came with Buckland. Milman tells me that
Buckland is very much liked, as Dean, in everything he has
done. We had another Geological Dean, W. Conybeare,
lately made Dean of Llandaff. He desired me to send you
his best regards, as, indeed, have many Knights of the
Hammer. Forbes, I am happy to say, is better ; so much so,
as to have been able to give a lecture at the Royal Institution
last Friday, notwithstanding the remonstrances of all his
friends, medical and others ; I hope he has not suffered. His
lecture was " Whence and where came the existing Flora and
Fauna of the British Isles." He could do little more than
touch upon the flora. He shews a Scandinavian flora on
the mountains of Scotland and Wales ; a Germanic one in the
middle regions of the East of England, and Norman in the
South and South-west, and certain plants of the South of
Ireland he traces to the Asturias, and believes that during the
Miocene period Spain and Ireland were united. He touched
upon some of his discoveries in his last summer's dredging,

and told us that, as he hoped to do, he had found in some *very deep* places in the Atlantic, off the North-east coast of Scotland, living shells of an undoubted Arctic character.

You will think that Geology excludes all other subjects from my mind. It is pretty nearly true that every hour unoccupied by factory work, is given to Geology; there is no other way in which I could get through my duties as President without certain failure.

God bless you,
Your affectionate father-in-law,
LEONARD HORNER.

To Charles Lyell, Esq.

London, 1st *April*, 1846.

MY DEAR CHARLES,—I am very sorry to begin this letter with a piece of intelligence which we received this day, and which will give you and dearest Mary much pain; there has been a great battle on the Sutlej, in which your brother's regiment was engaged, and he is reported to have been very severely wounded. I trust that he will speedily recover, but there must be great anxiety until more particulars are learned. They will be kept in a most painful state of suspense at Kinnordy. The victory is a very great one, gloriously won, and I trust may be attended with good effects, but until we hear whether the Sikhs will defend Lahore, we cannot be sure that there will not be more fighting. Darwin gave us a paper at the last meeting on the Falkland Islands, he was present and spoke, and we had an excellent discourse from Forbes on the distribution of species over wide areas.

Joseph Hooker is appointed botanist to the Geological Survey of Great Britain. We are to have a long paper by Murchison on the detritus of Scandinavia and the erratic blocks, the result of his travels last summer. What I have

read of it is good ; a little too diffuse, and too much of what
he has already given in his " Geology of Russia." Falconer
is going on steadily with his work. He told me on Monday
that he had clearly established that the Mammoth of Siberia
in the older diluvium, in the detritus that covers the east
flanks of the Urals, is a distinct species from those found on
the banks of the Lena, etc.

I dined with Hallam yesterday, who is very well. We had
Lord Lansdowne, Lord and Lady Monteagle, Edward and
Mrs. Romilly, Sir John and Lady Catharine Boileau, and the
Bishop of Oxford, Wilberforce ; an agreeable party, the great
battle in India the chief topic.

The Corn Bill still lingers in the House of Commons, and
there is no prospect of its being out of that House before the
holidays. The delay is causing much inconvenience and loss
in trade. There seems to be a strong feeling that Peel's
Government will break down shortly, because of the hatred
Protectionists bear towards him. I have no such expectation,
and as a well wisher to Liberal measures, I hope he will
remain. Unless by a general election it should appear that
there is an increased Whig feeling in the country to send a
good majority to Parliament, I have no expectation that the
Whigs will come in. If the dissolution be delayed for a year
my belief is that Peel will come in stronger than ever, and
with an increase of his power to carry Liberal measures. But
a dissolution now, would give rise to much asperity, so
irritated are the high Tory party by what they call Peel's
desertion of them. My love to my dearest Mary.

Ever, my dear Charles,

Affectionately yours,

LEONARD HORNER.

To his Daughter.

Manchester, *April 23rd*, 1846.

MY DEAREST SUSAN,—Mr. Hallam's remarks on the imperfect education of our artists is but too true; and I am not at all surprised at it. With rare exceptions they belong to the classes somewhat below the middle, and nothing can be worse than the educational opportunities in this country for anything like a liberal education for such persons. It is a branch of the subject of education which has never been sufficiently dwelt upon. Schools for the poor are much talked of, and some little has been done, but a collegiate education for the youth of the middle ranks seems to be quite lost sight of. If the young men are badly educated, I suppose it is quite as bad, if not worse, with their sisters. I was glad to hear Mr. Hallam's testimony to the superior accomplishments of Mr. Eastlake.

I hope you are getting on to your satisfaction with Joan of Arc.

I have absolutely nothing to tell you. I am harnessed to my work in the morning, and until about eight at night, and then I get some reading—chiefly matters of history that happened ten million years ago. My love to all.

Your affectionate father,

LEONARD HORNER.

To his Daughter, Mrs. Bunbury.

Manchester, *April 28th*, 1846.

MY DEAREST FRANCES,—I regret very much not having been present to hear Charles' discourse on Stigmariæ, which Moore tells me he did well, as I have no doubt was the case.

You appear to have had a very pleasant party at Sir James Clark's. Mr. Wyse is a very agreeable man. I hope he gave you some comfort about Ireland, for one needs it, so much there is of sadness in the condition of that country. Kew Gardens must be a great treat to all botanists, I regret very much that I never studied that branch of science; the physiology is so very interesting, and the objects themselves so beautiful. De Caudolle's " Organographie " is the only botanical work I have ever read, and I was charmed with it. I do not think it is too late to go to school for any worthy object, whatever a man's age may be; and there is certainly none more calculated to soothe the evening of life. I am very sorry for Sir Charles Napier's disappointment, had he come in time for the battle of Sobraon he would probably have been in the thickest of the fight, and his friends may therefore rejoice, for he does not require any fresh laurels.

Ever, my dearest Frances,

Your affectionate father,

LEONARD HORNER.

[Mr. Horner took an active part this year with Mr. (afterwards Sir William) Grove, in bringing about a much needed reform in the Royal Society, which was losing its scientific character, through the influx of an unrestricted number of Fellows being annually admitted, without reference to eminence in their qualifications. It was a hard struggle, but ultimately attended with success.]

To W. R. Grove, Esq.[*]

Manchester, 12*th May*, 1846.

MY DEAR SIR,—I enclose a copy of a letter and financial statement which I have sent by this post to Dr. Roget; and I have told him that I have sent you a copy.

You will see that our belief that there is no ground for opposing the limitation of new Fellows, on account of revenue, is fully justified by the examination I have made. If it was

°The Right Hon. Sir William Grove,

not giving you too much trouble I should be glad to hear the
result of your intended visit to Lord Northampton, and whether
you have found other Fellows entertaining the views we do,
besides those you mentioned to me.

Yours faithfully,
LEONARD HORNER.

To W. R. Grove, Esq.

Manchester, 16*th May*, 1846.

MY DEAR SIR,—So important a change in the constitution of
the Royal Society as is involved in our proposal, can hardly
be expected to meet with the assent of Lord Northampton and
many others, until it has been much talked of, and until a
strong feeling in its favour is manifested. I do not believe
that anything short of a limitation of the numbers, will place
the Royal Society in that high degree of estimation in which
it ought to be held, both at home and abroad. Neither in
your letter, nor in one I had from Dr. Roget, does there
appear any *reason* to have been assigned for not assenting to
the limitation, except the apprehension of financial difficulties,
which my calculations seemed to show there is no ground for;
and it does not appear that I have made any material
mistake. . . . I think it would be well to have the
question of limitation fairly discussed in the Council, and
that every member of it should be called upon to state his
reasons for and against it. Might it not be advisable that
some one should move, and another formally second, the
question "that it is desirable that the number of ordinary
Fellows should be limited." Upon this let the discussion be
raised.

Making the ballot more stringent is good as far as it goes,
but it will not effect the great object we aim at. It would be
a great mistake, I think, to ask for a new Charter, unless for

some great objects; it would postpone to a very distant
period any attempt at essential changes, which would require
another Charter.

> I am, my dear Sir,
>> Very truly yours,
>>> LEONARD HORNER.

To his Son-in-law, Charles Bunbury, Esq.

Combehurst,° *June 6th,* 1846.

MY DEAR CHARLES,—It gave me great pleasure to receive
your letter yesterday, and to learn that you and Frances are
having so much enjoyment. I wish, with all my heart, that
I could join you and then I could explore Cader-Idris
with such companions, with calm quiet leisure; but
that is impossible, this year at least. Your mother-in-
law, Susan, and Kate, and I, came here yesterday, to
dinner; Frances will tell you how very enjoyable a place it
is, and what very agreeable people its possessors are; then we
have glorious weather, very warm certainly, but I do not wish
for any change for a long time, so much good is to be done
and so much enjoyment to be had in this genuine summer
heat, which we so seldom have.

I am very much gratified by what you say about my Address,
for I am certain that you are too honest to say anything you
do not feel, and I hope you will give me equal credit for honesty
when I say that on no one's judgment of a philosophical
writing would I place greater reliance. The structure of the
country around you must be very interesting to examine, from
what you say. There was a paper by Aikin on Cader-Idris
long ago, with a good map of the mountain. It is in the
"Transactions of the Geological Society," and is one of the
memoirs to be had separately. I believe Ramsay to be an

° The residence of Samuel Smith, Esq.

excellent observer, and I should think, if you have an opportunity, a day or two with him, while he is at his surveying
work, would be very interesting and profitable.

Ever, my dear Charles,

Affectionately yours,

LEONARD HORNER.

To his Daughter.

Accrington, *July 25th*, 1846.

MY DEAREST KATHARINE,—I thank you for the very agreeable letter I have just received from you. You know that in
these sad absences from home, the post is my chief pleasure,
and that all the family details are most interesting to me. I
have no doubt that you are very happy at Hampstead,* I
never left that house with other than an agreeable impression,
and always go to it with pleasure. I am sorry that you do
not give me a better account of Mr. Mallet; very hot weather
does not agree with him. The addition of Mr. Smyth's†
society is a great gain, and it is a great privilege to see so
much of that truly excellent man.

It is very pleasant to see Mr. George Thomson enjoying so
happy an old age, and he has done right to go back to his
native place. His granddaughter married Dickens, and he
lived for some time in their neighbourhood. I heard of some
great amateur concert having taken place not long since at
Edinburgh, and that Thomson took his place in the orchestra.
Your evening readings must be made extremely interesting by
Mr. Smyth's commentaries.

Mr. Smyth does not like agitators; they are in general
persons who must disturb both his natural temperament and
the calm equanimity of his years, but there is this great exception in the case of Cobden, that he agitated not to gratify

On a visit to Mr. and Mrs. Mallet.

† Professor Smyth.

personal ambition, but to rouse the public mind to a lively
sense of the consequences of an injurious system of legislation.
In his addresses, he was sometimes, in unguarded moments,
hurried into unjustifiable expressions, but he is a very fair,
candid man, of great practical sagacity and singleness of
purpose. He has earned for himself a very distinguished
place among the leading men of this country. Whether he is
fitted for official life is another story. He would be thrown
into a society and sphere so totally new, that it is not easy to
predict how he would show himself fitted for them. I see that
he has announced his intention of retiring from public life
for a year. He will at the end of that time, if his health
permits, be in such a situation as to money, as will enable
him to choose his line—I speak of the fund now raising for
him.

To his Daughter.

Bedford Place, *23rd August,* 1846.

MY DEAREST FRANCES,—I assure you that I very often
think of the happy three weeks I passed at Mildenhall, and
the pretty lawn and house are very often in my mind's eye.

I have contrived to spend from two to three hours most
days in the British Museum, following up my Egyptian
researches. I have been entirely engaged hitherto with the
great French Work. It is a mass of most valuable information upon that country, of all sorts, collected by very able men
with great industry and talent.

I went with your Uncle Lloyd to the Zoological Gardens.
I had not been there for three or four years, and I of course
saw a great change. The gardens are quite beautiful, and
there is a noble collection of healthy animals. Your Uncle is
dining with us, and I am now writing between dinner and tea.

I saw Richard Griffith, the geologist from Dublin, on
Friday. He has been appointed Vice-President of the Board

of Works there, and is come over to give information to
Government about the state of the people, and the means of
applying the intended relief, called for by the failure of the
potato crop. He has been engaged in the same work since
last autumn. He told me that the disease is universal
throughout Ireland, that *last* year there was a very large crop,
that about one half was lost for human food, but as a part
was food for pigs, on which they fattened, it may be said that
the total lost did not exceed one-third of an unusually large
crop. But this year matters are much worse : in the first
place, not half the usual quantity was planted, and it is not
expected that more than one-fourth of that half quantity will
be fit for food to man or beast, and even that fourth to be
saved must be consumed immediately. He says that from
the measures Government have taken, he is not afraid of
there being sufficient supply of Indian corn to make up the
deficiency, but there is the greatest possible difficulty to give
the relief, and not do mischief by giving encouragement to the
idle habits of the people. Much harm has been already done
by the relief that was afforded since last autumn. He says
that there are at least two millions of people who have been
in the habit of subsisting entirely on potatoes.

The appointments by the Government have given satisfaction, but he says that Ireland never will be without agitation
until the Catholic priests are paid by Government, and I
believe he is right.

From Mr. Charles Darwin.

Down, Farnborough, Kent, *August 10th*, 1846.

MY DEAR MR. HORNER,—In following your suggestion in
drawing out something about Glen Roy for the Geological
Committee, I have been completely puzzled how to do it. I
have written down what I should *say*, if I had to meet the
head of the Survey, and wished to persuade him to undertake

the task, but as I have written it, it is too long and ill expressed, seems as if it came from nobody and was going to nobody, and therefore I send it to you in despair, and beg you to turn the subject in your mind.   I feel a conviction if it goes through the Geological part of Ordinance Survey, it will be swamped, and as it is a case for more accurate measurements, it might, I think, without offence, go to the head of the real Surveyors.   If Agassiz or Buckland are on the Committee they will sneer at the whole thing and declare the beaches are those of a glacier lake, than which I am sure I could convince you that there never was a more futile theory.

I look forward to Southampton with much interest, and hope to hear to-morrow that the lodgings are secured to us.

You cannot think how thoroughly I enjoyed our Geological talks, and the pleasure of seeing Mrs. Horner and yourself here.

Ever your obliged,
CHARLES DARWIN.

To his Daughter.

Manchester, *October 22nd*, 1846.

MY DEAREST KATHARINE,—I went to Bury yesterday and in the forenoon called at the Rectory* where I found all the inmates at home and well.   Yesterday I dined at Sir Benjamin Heywood's, and it was an agreeable party.   We had Lord Morpeth, whom I had not met with before, and whom I liked during the brief opportunity that a dinner gives; his brother, Mr. Charles Howard, whom I had a long talk with, Lady Mary Howard, Sir Thomas Arbuthnot, the commander of the forces in this district; his Aide de Camp, Mr. Fane, a pleasant looking young man; and George Loch and his wife.

° The Rev. Geoffrey Hornby's.

I mentioned to your Mamma that I had had an excellent
and remarkable letter from Florence Nightingale, commenting
on Mr. Braid's pamphlet, and maintaining that his deductions
and reasonings are in the main the same as the mesmerists,
whose doctrines, as held by Dr. Howe (of Philadelphia, I
believe), she appears to adopt. I called on Mr. Braid and
left him the letter, and last night I received a long reply from
him to Florence's observations, which I shall send to her;
when I get it back I shall send you both, and it will be a good
exercise of your metaphysical powers to study them along
with the pamphlet. It is a curious *psychological* question, but
it is of great importance to be thoroughly investigated, if, as
appears to be made out in some cases, it is susceptible of
curing or alleviating bodily suffering.

To-day I am going to Patricroft to dine with Mr. Nasmyth
and see his machinery.

Love to all, dearest Kate,

Your affectionate father,

LEONARD HORNER.

To his Wife.

Manchester, *October* 25th, 1846.

On Friday I went to Patricroft at three. Mr. Nasmyth
took me over his works, and explained the steam hammer to
me, and many other objects of great interest in his way. I
saw several of these wonderful hammers making, one for
Cairo, another for Stockholm, and another for Berlin; and more
than one for this country. He had sent off one the day before,
for the copper and gold works at Ekaterinburg on the Asiatic
side of the Urals. One that he is now making for Glasgow,
to forge the great iron beams of engines for steam vessels, is
immense; the hammer that falls weighs *six tons*, yet so
completely under control is this vast power, that he told me
he would place an egg in a wine glass under it and he could

let the hammer fall with such gentleness that it would be no
more than breaking the egg shell like the tap of a teaspoon.
In one he uses at the works, I saw a nail driven into a board
with a succession of gentle blows as you would do with a light
hand hammer, and the next minute it welded a vast mass of
iron at a white heat as a baker would knead a mass of dough ;
the weight of this is one and a half tons.   I then, at five
o'clock, went to dine with him.   He lives with great simplicity
in a very small house near his works, there was no one there
but his wife, a very sensible person, the daughter of an iron-
master at Rotherham.   I never passed a more agreeable
afternoon.   He is full of genius in a variety of ways ; he has
made two very large telescopes which stand in his garden,
with which he has been for years observing the moon, and
he has made a series of most curious drawings, exact pictures,
and has some curious speculations about the volcanic structure
of the moon.   Then I looked over some portfolios of drawings,
sketches, etc., taken in Italy, Germany and Sweden, and some
beautiful oil paintings, all by himself.   Then there is an
unpretending simplicity about him and an enthusiasm that
are very agreeable to find.   It was a great pleasure to me to
see this excellent man at the head of a great establishment of
six hundred and fifty men, with an European reputation, and
making a large income, whom I knew as an ingenious boy,
and who struggled against the greatest difficulties in youth.
I could dwell also on some excellent traits in his character,
but these I reserve.   I came home by a train at ten o'clock.

———

To Charles Bunbury, Esq.

Manchester, *6th November*, 1846.

MY DEAR CHARLES,—At breakfast this morning in Bedford
Place, I had the pleasure to receive your letter of yesterday.
I left Euston Square at ten, and was here at five.   I had the

good luck to have the Bishop of Norwich in the same carriage, and the day was very fine, a day worthy of *le petit été de St· Martin.*

The important Council of the Royal Society, which I looked forward to with some anxiety and much interest, when the report of the Charter Committee was to be discussed, took place yesterday. You will be glad to hear that the side on which I fought gained the day. You are aware that the great object of those with whom I act, is to limit the number of fellows to be elected in any one year, to fifteen, and that the most eligible among the candidates proposed, shall be recommended to the Society for election by the Council. This proposal has been most strenuously opposed by Lord Northampton; he did not come yesterday, but we met sixteen, only three others being absent. Sir John Lubbock proposed that the consideration of our Report should be adjourned *sine die,* and was seconded by the Dean of Ely, and supported by Professor Willis, the two Secretaries (Roget and Christie), and Mr. Galloway. Ten voted on our side, Rennie (in the chair), Colonel Sabine (Foreign Secretary), Smyth, Wheatstone, Daubeny, Grove, Royle, Sharpey, John Taylor, and myself. That motion being so disposed of, the Council voted their approbation of the two leading principles of the recommendation of the Committee (those above mentioned) and appointed a committee to take measures towards a further prosecution of the recommendations of the Charter Committee. This is a great triumph and the commencement, I hope, of a better state of things in that Society.

The article in the *Quarterly* on education *is* by Milman. I talked with him about it, and he admitted his being the author. I told him that he had rendered a great service to the cause, and had greatly smoothed the way for Lord John. Much as that great step of the Government is wanted, I wish that it could be postponed until after the next election, for anything that will satisfy the Dissenters will be objected to by the church, and *vice versa*,

My kindest love to my dear Frances.

Your affectionate father-in-law,

LEONARD HORNER.

———

To W. R. Grove, Esq.

Manchester, *November 8th*, 1846.

MY DEAR SIR,—I wrote to Colonel Sabine, suggesting that the Committee for preparing the case for the opinion of Counsel should meet very soon, as it appears to me most desirable that we should have that opinion before the Anniversary.

I hope that you will have a great deal to do with the preparation of the case; I am sure you will bestow every care upon it. It appears to me very important that we should not confine ourselves to asking the question, whether the Council have power to limit the number to be elected in any one year, but should set forth the mode in which it is proposed that the election shall take place, in order that Counsel may see that the utmost care is taken to preserve every right and privilege of election conferred by the Charter on the Fellows; which right and privileges each will have power to exercise uncontrolled to the extent of fifteen persons. That seems to be the only question to be asked, for there can be no doubt of the Council having power to recommend, that is, express their opinion of the comparative eligibility of the candidates, as well as to construct the machinery by which the intended measure is to be carried into effect. It is possible that the power of limiting the election to one day may not be so clear, but I have neither the Charter nor Statutes here.

I may have said nothing in this note that would not certainly have occurred to yourself, but I know you will excuse

my giving utterance to my thoughts, for the prospect of good
that is now likely to arise to the R.S. is often before me here
in my solitude.

Yours faithfully,

LEONARD HORNER.

To Charles J. F. Bunbury, Esq.

Bedford Place, *20th December*, 1846.

MY DEAR CHARLES,—You asked me to give you some account
of the proceedings at the Geological Society last Wednesday.
Sedgwick came to town with a very bad cold, so he took the
field with some disadvantage. He brought a very long paper,
with numerous sections, part of which only he read, delivering
the rest he had to say orally, and he gave us an hour and a
quarter of reading and speaking.

He had some descriptions of parts of North Wales he visited
last summer, but the chief points in his communication was
to stand up for the restoration of the lowest sedimentary rocks
into a distinct system, with his old name of Cambrian, and
not to have them merged into the Lower Silurian beds. His
main arguments are, distinct lithological characters, position
and *great development*, and throwing zoological evidence into sub-
ordinate importance. He was answered most ably by
Murchison, who clearly established the superiority of zoological
evidence as a basis of classification above all other, and as
Sedgwick had claimed certain rocks as Cambrian, which
Murchison had classed as Lower Silurian, and as he even
went so far as almost to deny the application of the term
Silurian to any other than the upper part of the series, Forbes
shewed that in a deposit near Bala, in the very heart of
Sedgwick's Cambrian Rocks, numerous specimens of fossils
had been found, that are common both to the Upper and
Lower Silurian beds in other places. If we had come to a

division among those geologists present who understand the
subject, I suspect that the learned Professor would have gone
forth alone, notwithstanding his eloquence and his long established authority.

My kindest love to Frances.

Yours affectionately,

LEONARD HORNER.

From Mr. Charles Darwin.

Down, Farnborough, Kent, 1846.

MY DEAR MR. HORNER.—I am truly pleased at your approval
of my book, and it was very kind of you taking the trouble to
tell me so. I long hesitated whether I would publish it or
not, and now that I have done so at a good cost of trouble, it
is indeed highly satisfactory to think that my labour has not
been quite thrown away.

I entirely acquiesce in your criticism on my calling the
Pampean formation "recent"; pleistocene would have been
far better. I object, however, altogether on principle (whether
I have always followed my principle is another question) to
designate any epoch after man. It breaks through all
principles of classification to take one Mammifer as an epoch.
And this is presupposing we know something of the introduction of man : how few years ago, all beds earlier than the
pleistocene were characterized as being before the monkey
epoch. It appears to me that it may often be convenient to
speak of an Historical or Human deposit in the same way as
we speak of an Elephant bed, but that to apply it to an epoch
is unsound.

I have expressed myself very ill, and I am not very sure
that my notions are very clear on this subject, except that I

know that I have often been made wrath (even by Lyell), at
the confidence with which people speak of the introduction of
man, as if they had seen him walk on the stage, and as if in
a geologico-chronological sense it was more important than the
entry of any other Mammifer.

You ask me to do a most puzzling thing, to point out what
is newest in my volume, and I found myself incapable of doing
almost the same for Lyell.   My mind goes from point to
point without deciding; what has interested oneself or given
most trouble is, perhaps, quite falsely thought newest.   The
elevation of the land is perhaps more carefully treated than
any other subject, but it cannot, of `course, be called new.   I
have made out a sort of index, which will not take you a
couple of minutes to skim over, and then you will perhaps
judge what seems newest.   The summary at the end of the
book would also serve the same purpose.

I do not know where Elie de Beaumont has lately put forth
on the recent elevation of the Cordillera.   He "rapported"
favourably on d'Orbigny, who in late times fires off a most
royal salute, every volcano bursting forth in the Andes at the
same time with their elevation; the debacle thus caused,
depositing all the Pampean mud and all the Patagonian
shingle! Is not this making Geology nice and simple for
beginners?

With many thanks, most truly yours,

CHARLES DARWIN.

P.S.—I am astonished that you should have had the
courage to go through my book.   It is quite obvious that
most geologists find it far easier to write than to read a
book.

Chapter I. and II.   *Elevation of the land*, equability on East
coast, as shown by terrace, page 19; length on West coast,

page 53 ; height at Valparaiso, page 32 ; number of periods
of rest at Coquimbo, page 49 ; elevation within human period
near Lima greater than elsewhere observed, the discussion,
page 41; on more horizontality of terraces, perhaps one of the
newest features ; on formation of terraces rather newish.
Chapter III., page 62. Argument of horizontal elevation
of Cordillera, I believe new. I think the connection (page 54),
between earthquake starts and insensible rising, important.

CHAPTER VI.

1847—1848.

To Charles Bunbury, Esq.

Bedford Place, *27th February*, 1847.

MY DEAR CHARLES,—I am happy to find that you are making progress in the recovery of your strength. I shall be glad to hear that you are in Torquay, sunning yourself in Frying-pan Row, the sheltered side of the harbour. I wish I could join you, for we might on a mild day have some very nice little bits of geology near at hand to feast upon. Watch for mild days to visit Babbacombe Bay, where there is much to interest you; and you will not fail to notice the groves of fuci, waving erect in the clear aqua-marine.

We had an excellent meeting last Wednesday. It was entirely taken up with a valuable paper by Prestwich, the result of twelve years' observation, making out very clearly that the so-called London clay of Brackle sham near Portsea and of Barton in Hordwell Cliff, is not the equivalent of the London clay in the London Basin, but of posterior date, and that the *Calcaire Grossier* of the Paris Basin, supposed to be synchromous with the London Basin, is the equivalent of the Hampshire beds. This view involves a very different state of the sea bottom over that area at two different periods. The paper was much lauded by all who spoke, and the observations it gave rise to, lasted till eleven o'clock.

I leave London for Manchester on the 25th. I hope there is a probability of your being in town before that day. I am very sorry that you will not be at our anniversary. I am

pretty well advanced in my Address. If I could have given
continued attention to it, the preparation would have been a
source of pleasure; but to execute such a work by snatches of
time, and in the evening only, after the day's work, is both
unsatisfactory and laborious.

Yours, my dear Charles,

Very affectionately,

LEONARD HORNER.

To his Daughter.

Manchester, 12th *March*, 1847.

I thank you, my dearest Frances, for your affectionate
letter.

I had heard from your letters to Bedford Place that you
were both well pleased with Torquay, but what you tell me of
the great object of your going there, the improvement of your
husband's health, I am'particularly glad to hear. It is indeed
a most charming place, both for climate, sea views, and land
views. If Charles is already able to walk as far as Babbacombe, he will soon be able for some geological investigations,
and as there are donkey cars, you can always be with him
without fatigue.

I have been much pleased with several articles in the last
number of the *Edinburgh Review*, but above all with that on
Pascal, which I agree with you is quite admirable. I shall
read it again soon, and shall also read the book reviewed. I
knew comparatively little of Pascal's merits before. I did not
find the article on Hume too long, but I rose from the perusal
of it with a lower estimate of Hume's character than I had
before; for he appears to have been under the influence of
strong prejudices very unworthy of a philosopher, and which
are calculated to shake one's confidence in him as such. I
should like to know who wrote the article on the streets of

Paris; it is very diverting, and shews an extraordinary acquaintance with the social history of Paris for many centuries.
I am leading a very regular life, and am systematically, as far
as I can, confining my official work between 9 a.m. and
5 p.m., and I have the pleasure of passing my evenings very
agreeably with my books, partly geological, partly general
reading.

I have not seen anyone except Sir Benjamin Heywood, for a
very short time, and Sir Thomas Arbuthnot, with whom I sat
an hour and a half on Sunday, talking over many things. He
is commander of the district, and his command ranges over
seventeen counties. He receives regular reports from the
officers subject to his command, and these he makes a monthly
digest of, and forwards to Sir George Grey and the Duke of
Wellington, to keep them acquainted with the state of the
manufacturing population, which in these times of dull trade
is very important.

My kind love to Charles, my very dear Frances,

Your affectionate father,

LEONARD HORNER.

[At the end of May, Mr. Horner was far from well, and Sir
James Clark advised rest from all work for a time, it was
therefore arranged for him to take his annual holiday earlier,
and he and Mrs. Horner and one daughter went to Germany,
where one of his daughters was already spending some months
with a friend.]

To his Daughter.

Bonn, *June* 10*th*,1847.

MY DEAREST KATHARINE,—I intended to have written to
you at some length to-day, but since I sat down, Mr. Erskine,
Dr. Mendelssohn, and Carl Brandis, have been here, and we
are going out in two hours to Rolandseck. After a due

consideration of all circumstances, I have resolved to extend
our journey, so as to see a little more of true Germany than
we have ever yet done. We start on Saturday by the steamer,
and shall visit the Holwegs at Rheineck. Our grand aim is
Nürnberg! To be so near you (at Kissingen) and your dear
kind companion, and not to visit you is impossible, *coute qui
coute*. We hope to get to Aschaffenburg on the 15th and to
Kissingen next night.

I have been enjoying myself greatly here, and although I
have had a slight return of the affection of my head, I am
better, and my legs have not given me any trouble since I
left England, and yet I have used them a great deal.

My very dear Kate,

Your affectionate father,

LEONARD HORNER.

————

To his Daughter.

Donauwerth, *June 20th*, 1847.

MY DEAREST KATHARINE,— . . . We saw Nürnberg
thoroughly, and were much pleased with it. It more than
answered all our expectations. I will send you, before you
leave Kissingen, a short account I bought and read of the
battle between Gustavus Adolphus and Wallenstein, close by
Nürnberg, in 1632, and it was most interesting to look upon
the very spot where it took place.

We started at eleven last night. The evening was fine but
cloudy, and rain very soon came on, which lasted the whole
night, and till one o'clock to-day. Our road lay by Weissenburg
and Manheim. Just as we entered the little town of Ellingen
at four in the morning, by the breaking of the axle the
Eilwagen was overset. Thank God! none of us were hurt.
I got a slight blow on my head, but it is of no consequence.
. . . . . . A neighbouring doctor, near to whose house

we fell, heard the crash and came out, and most kindly
received us into his house. Soon after we got into a Gasthof
and had some coffee, and the Conducteur having found a
chaise, we got into it and proceeded to a station beyond
Weissenburg, where we found a spare Eilwagen. We propose
proceeding by a railway train from here, and hope to reach
Munich to-night soon after nine.

I am, my dearest Kate,

Your affectionate father,

LEONARD HORNER.

To his Daughter.

Coblentz, *June* 30*th*, 1847.

MY DEAREST KATE,—Our last despatch was from Stuttgardt,
which we left for Heilbronn, and as I had learned that of the
two Eilwagens that go, one at six a.m., the other at ten, the
earlier passed through Marbach, we had no hesitation in
preferring a sight of Schiller's house to a longer sleep, and
went that way. We saw the house, now a baker's shop, and
over it is a sign " In diesem Hause ist geboren Friederich von
Schiller der 10th November, 1759." There are no records
respecting him, but we inscribed our names in a book kept for
that purpose by the baker. All we got from him were some
rolls, which he called " Schiller's Brod " and which we found
to be like his works, most excellent.

We passed through a pretty country, and got to Heilbronn
at mid-day and went to afternoon service, where we heard a
most capital sermon on the Reformation and on *Glaubens
Freiheit*, for it was the first Sunday after the 370th anniversary
of the signing of the Confession of Augsburg, and there is
annually a sermon to commemorate that great event. The
sentiments of the preacher were admirable, but I cannot say
much either for the purity of his dialect, which was *echt*

Swäbisch, or for the euphony of his name, which was *Pfarrer·
Buttersack.*

We came down the Neckar from Heilbron to Heidelberg,
a most beautiful valley.

Your affectionate father,

LEONARD HORNER.

To his Daughter.

Burnley, Lancashire, *July 18th*, 1847.

MY DEAREST KATHARINE,—I often and often think of the
very happy day we spent at Kissingen, and of our evening
walk in that pretty wood.   I was rejoiced to learn by your
letter, that Maria* had derived so much benefit from her stay
at Kissingen, and that account was confirmed to me yesterday
at Barlow.   I fancy you now in the Baierische Hof in Munich,
and ever since Monday I have been travelling with you.   It is
so pleasant to know the road you have been passing over, but
alas! you go to-morrow into a land as yet unknown to me,
and I fear likely to remain so.   I take it for granted that you
have had a pretty detailed account of our journey.   I look
back to it with the greatest pleasure, and every part has left
a vivid impression.   All was interesting, but Nürnberg and
Munich were the great features, and the day's journey between
Ulm and Stuttgart, through so pretty a country, and the
Neckar from Heilbronn and Heidelberg.   Be sure to drive to
the Wolf's Brunnen, and along the upper road from thence to
the castle.   We passed a most agreeable evening with the
Mendelssohns at Horcheim, and three very pleasant hours at
Rheineck† the next day, lamenting as you and Maria did,
that we could not sojourn longer with that excellent family.
Then Bonn, notwithstanding all we had seen, looking more
beautiful than ever, and we had a delightful evening at

* Miss Phillips of Barlow Hall.

† Mr. Bethman Holweg's lovely castle.

Rolandseck and saw much of Professor Brandis. I should be sorry indeed if I thought that I could not see Bonn and our dear friends there again, and before very long. My health has certainly benefited by the journey, and I shall take care not to over-exert myself in anything.

Well, we got safe home on the 6th, the day I had calculated upon, when we started with our face homeward from Munich, but not without an adventure, for we went ashore at the South Foreland in a fog in the middle of the day. Happily we had been going at half speed for some time, so that the shock was slight, and we got easily off without damage.

We had the happiness to find dearest Susan and Nora quite well. I could not get through my arrears of business preparatory to setting out in my circuit before the 13th; the most important event in the interval, was the visit of Hans Christian Andersen last Sunday at breakfast, with whom we were all pleased. I have here Andersen's "Märchen," they are many of them very amusing, and the quaint simplicity of the style pleases me very much; I have besides a volume of Niebuhr's "Lebensnachrichten," Prescott's "Conquest of Mexico," and Mr. Tourguenef's *Russie et les Russes*, so that I have no want of amusement before breakfast, and in the evenings. I am up at six, and shall continue that good practice as long as I can.

———

[The lease of Mr. Horner's house in London having expired, he removed with his family to Rivermede, in Hampton Wick, near Kingston-on-Thames.]

———

To his Wife.

Lancaster, *September* 25th, 1847.

This day, my dearest, dearest wife, I got the first letter you have written to me from Rivermede.

Now that the bustle and occupation of removal are over,

and that we are fixed in our new abode, one has time to reflect upon the change. It is an epoch in the evening *of our* lives, a change of mode of life, as well as of place. We shall have to adapt ourselves to the altered circumstances ; and of these the greatest is the comparative stillness, the absence of the excitement which the short morning visits of friends creates, the withdrawal, to a certain extent, from the bustle of passing events. This will fall with greater force on the younger minds of our dear daughters, but I have no fear of their pining after what they have left, for they will not fail, I am sure, to find out many sources of happiness.

My thoughts are with you and them all day, and I go over the house, and through the garden, and see all the pretty trees on the lawn, and thus keep myself at home. I look forward with confidence to our being very happy in our new mode of life, should it please God to keep us all in health.

As my work ends at five o'clock, I have time for much reading, and thus my time passes as cheerfully as it can do, away from home and alone.

I am much pleased with the " Dichter's Bazaar "; Andersen certainly has the rare talent of making the most ordinary events of life interesting. I read this evening his description of Nürnberg and Munich, and was transported back to the charming days we so lately passed in those places. I never made a journey that has left a more vivid or more agreeable impression upon me. Andersen does not do justice to Munich I think, he is charmed with Nürnberg. In many respects alike, and no less a master of description, is Hugh Miller, whose " First Impressions of England " is one of my books. His mind is a more powerful one than Andersen's.

Two copies of Mr. Dawes'* pamphlet were sent me. I do not know by whom either was sent. I have one with me, but have not yet read it. I shall do so with more interest from

° Afterwards Dean of Hereford.

what you say of it. I enclose an excellent letter I have had from John Poole of Enmore. I am glad you had such good weather for your first walk through Bushey Park.

To his Daughter.

Kendal, *September 27th*, 1847.

MY DEAREST KATE,—After taking coffee, before going to bed, I read Mr. Dawes' pamphlet. It is excellent throughout, full of practical good sense, enlightened sound views as to what is wanted to place schools for the great mass of the people on a proper footing, and breathing throughout benevolence and respect for his humble fellow creatures I especially admire his views as to the bringing the children of the employer and the employed into the same school ; the effect must be to humanize both, and create a right feeling between them. We saw this carried into effect with entire success at Mr. Poole's school at Enmore, in 1813, as your Mamma will remember ; and if I mistake not, you will find in one of your Uncle's letters in the autumn of that year, some remarks on this subject, for he visited the school at our suggestion, with William Adam, afterwards the Accountant-General, they being then on the circuit. I shall do all I can to make Mr. Dawes' pamphlet known in my district.

It was a beautiful evening yesterday, a clear, deep, blue sky, with the almost full moon brilliantly splendid. It has been a no less charming day, the sky without a cloud, and I have had great enjoyment of it. I set out at half past eight to visit mills at Beck, Mealbank, Burneside, Slaveley and Crook. They are all woollen mills driven by water, and most of them in pretty situations ; that at Mealbank in particular, in a deep bosky dell, watered by a clear, bright, swift running river, the Mint. It is a large new establishment, in the best order, erected by Mr. Braithwaite, a Quaker ; but he has not thought

only of having a good well-ordered mill; he has thought of
his workpeople, built excellent houses for them, which are
models of cleanliness, and an excellent school, in which I
found a large number of the cleanest, freshest children,
receiving excellent instruction from a schoolmistress educated
at the Home and Colonial School in Gray's Inn Road. I
found in the school a very good and apparently well-selected
lending library. The overlooker of the mill asked me to go to
see the man who occupied the same situation when I was here
last, four years ago, now bed-ridden from paralysis, as he
thought it would please him. I found the poor old man in a
most clean bed, in a well-furnished room, and had a short
talk with him. I was out six hours, and as I was in an open
jaunting car, I had a famous blow of fresh healthy air. I am
going a similar round to-morrow, but the country is not quite
so pretty as where I was to-day. I read before starting, four
Odes of Horace, since I came back and after dinner I have
read some chapters of the "Dichter's Bazaar," part of his
descriptions of Rome, very graphic and very interesting. His
journey with the Vetturino from Florence to Rome is very
amusing, but he must surely caricature the selfish Englander,
for there could hardly be so odious a creature of any nation.
I shall pass a part of my evening with Hugh Miller, and now
I will say good-bye for the present. I fancy you and dear
Mamma in our pretty drawing-room.

To his Daughter.

Carlisle, 30*th September*, 1847.

MY DEAREST NORA,—I daresay you passed your time very
pleasantly at Down.* He is a very superior being in many
ways, and in no man's society have I more enjoyment. I
regret much that I can see so little of him.

°At Mr. Charles Darwin's.

I have finished this evening two of my books, the second volume of the "Dichter's Bazaar," and Hugh Miller's "First Impressions." Andersen is a most graphic describer, and his book is exceedingly attractive and amusing, and my interest in his writings is much increased, since I saw and conversed with the author. But he skims along the surface, not so Hugh Miller, he places his foot on his spade, and turns up the ground wherever he goes. It is a most curious book, and shews the author to be a very remarkable mixture of two very different states of mind, free and unfettered in all subjects but one, and believing himself to be as free in his religious views as in his scientific, while it is evident that he is under the control of a fixed one-sided view on that subject. His seventeenth chapter, in which he deals with the Mosaic geologists, is admirable, and his contrasts of the English and Scotch characters in the two last chapters, are striking and original, and to a great extent very true, I believe. There is one blot in them, which I regret to see, his prejudice about the new Poor-Law system. He has a sneering way of talking of *Whig* measures, which is unworthy of him. His descriptions of Hageley, of the Leasowes, and of Olney, are wearisomely tedious. No man can give an idea of landscape by description in its minute details; the broad features may be told, but these only make an impression, create a picture before us, it is impossible to carry on the recollection of minute parts and hang them together.

I am very sorry to hear of poor Sir Edmund and Lady Head going to so cold a climate, with so very long a winter.

Love to dearest Mamma, and all your sisters. I hope Susan and Joanna are to return from Waverley to-morrow.

I am, my dearest Nora, your affectionate father,
LEONARD HORNER.

1848.

[Mr. Horner's third daughter, Susan, had gone abroad to
Italy with her sister and Mr. Bunbury, and in January his
fourth daughter, Katharine, was married to Captain Lyell of
the Bengal Army, brother of Mr. Charles Lyell.]

To his Daughter, Mrs. Henry Lyell.

Rivermede, *February* 10th, 1848.

MY DEAREST KATHARINE,—I have just come up to my study
after listening to a most delightful letter which the post
brought to me from dearest Susan. She and dearest Frances
and Charles appear to be enjoying themselves greatly, not
only among the many objects of interest which Genoa itself
affords, but having the additional joyous excitement which
the present political crisis in Italy is so calculated to call
forth. All we hear from Naples and Sicily appears to lead us
with some confidence to the expectation that a sober rational
freedom, if not all that could be wished, will be *planted,* at
least, with so good a root, and in so good a soil, that the tree
will flourish and quickly grow to maturity. If constitutional
liberty be established from Calabria and Sicily to the northern
frontiers of Tuscany and Sardinia, the iron despotism of
Austria in Lombardy cannot be sustained. It is evident that
a deep feeling of their degradation must have been long and
widely felt throughout Italy, when the spark struck out by
the enlightened Pope has kindled so quickly into a flame.

The correspondents in this house are so able and so active,
that I do not doubt you are kept in full possession of all that
is going on, and what a blessing it is, such a facility of
writing! how it keeps affections warm, and mitigates the
sorrows of separation.

Last week was a busy one in town—Monday, dining with
Grove; Wednesday, at the Geological Society Club, and the
Society in the evening; Friday, dining at Barlow's and Charles's

lecture—I was glad to get to the quiet of Rivermede on Saturday, for the excitement was too great. The paper at the Society was most interesting, on the bones of the gigantic bird from New Zealand.

I have only been in town on Monday this week, when I came home to dinner, bringing Lady Bell with me. I brought Factory work, which has kept my conscience easy, for I have given a fair portion of each day to the Queen's service; but I have enjoyed the calm quiet very much. The weather for twelve days has been mild and enjoyable. Your Mamma and I walked to Hampton Court to-day, and returned by the Home Park, quite delightful, the air so balmy, the deer enjoying themselves, three hares crossing our path, and the birds in full chorus. By-the-bye, I have made a discovery of two cork trees in the Home Park; did you ever see them? None here had done so.

I have finished Caroline Von Wolzogen's "Life of Schiller," and the interest of it was sustained to the last. It quite merits all the praise you gave it, and makes me admire the subject of it more than ever. I am going to read Carlyle's "Life of Schiller," and I believe I shall go through his works. You will think I am seized with a *Schillersucht;* if so, it was you who recently inoculated me. I have not made much progress in my Egyptian researches in the last fortnight, but I have kept them in sight.

To-morrow morning your Mamma and I go to town. I intend to be at the anniversary of the Astronomical Society to hear Sir John Herschel's discourse.

Your affectionate father,
Leonard Horner.

To Charles Bunbury, Esq.

Rivermede, *May* 15th, 1848.

My Dear Charles,—I have been very long in replying to your long and very interesting letter from Rome, of the 14th

April. It is a great happiness that you have all kept in such good health, and amid the turmoils of Italy, that you have been free from personal annoyance.

I am now sitting in my study, it is a quarter past nine, *Greenwich time*, both my windows are wide open, there is a cloudless sky, a bright moon, and a balmy air scented by the honeysuckle that covers the verandah at the porch-room door. In three-quarters of an hour, if he is true to his time this night as he has been hitherto, a nightingale will begin his delicious song; we have been out every evening for a week between ten and eleven listening to the dear fellow, who is the most indefatigable of lovers, carrying on his love song till two or three in the morning. The foliage is in the most exquisite state of verdure, and our lawn smooth and covered with daisies, and each day some new flowers appear. You know how much we were pleased with our new residence in the cheerless month of November, and may imagine how much more we are delighted with it now. The view from Kingston Bridge this afternoon as I returned home, both up and down the river, was beautiful. We have had a fortnight of real summer weather, the sun too powerful for the season, for it has burnt up the tender blossoms, and with this, and want of rain, we shall have no cherries. For three months there was continued rain; on the 26th of April the river had overflowed up to the plane tree on the lawn, Bully's grave was covered by the inundation; since then the surface of the Thames has fallen three feet, and we gardeners are praying for rain again.

Charles and Mary left us this afternoon—they came on Friday. Lord Cockburn came to us on Thursday, in time enough to see Hampton Court Gardens and the Home Park before dinner, and next day Combe Wood, Richmond Park and Bushy Park, he could not stay beyond Saturday morning; he was to set out this morning from London, and most

probably is at this moment within twenty miles of Edinburgh,
and probably will sleep at Bonaly to-night. When he came
to town, he left Edinburgh after a late breakfast at ten o'clock,
and was in Wilton Crescent soon after eleven at night. He
was remarkably well, and in excellent spirits.

To touch upon French politics would be dangerous, for it
would be difficult to know where to begin and where to end.
I rejoice that the retrograde Government of Louis Philippe
and Guizot has been put an end to, but I have no expectation
that a Republican Government can be fairly established or be
continued, in a country so little prepared by self-government
in internal affairs, with a people so restless and impatient, so
vain and self-satisfied. An assembly of 900, containing at
least 750 new men, wholly inexperienced in the rules by which
a deliberate assembly must proceed, and containing at least
850 persons who will insist upon talking or shouting, cannot
work to any good purpose. That so great a country as France,
with such vast natural resources, can fall back, is not to be
feared, but it will probably go through a long probationary
state before a Government suited to the nation is again established. The early years of Louis Philippe's reign were most
promising, and he is an eminent example what a corrupting
occupation royalty is. The fermentation in Germany is
immense, from one end to the other. A sensible, well-informed
Hanoverian, M. Von Lassert, introduced to us by Mrs. Brandis
of Bonn, dined with us to-day, and gave us a great deal of
information as to what is going on, and if what he expects be
realized, it will not be a regeneration of Germany only, but a
vast advance in freedom, that will not be confined within such
limits. He says that the hopes are, that all Germany will be
united under one Emperor, with responsible ministers, with
one army, one navy, one code of Custom-house laws, and *no*
State religion, every form having equal freedom, and one
supreme court to decide upon all questions between one state

and another. The seat of Government to be at a town in the
heart of Germany, and Erfurt is spoken of. The existing
Governments to continue, but modified, for the internal regulation of their several people's affairs, something like the State
Governments in North America. This is indeed a mighty
revolution, and God prosper it. M. Von Lassert says that
the opposition to be feared is not that of the Princes, but of
the wild young Germans, and the mob, who are wildly set
upon democratic rule. The great General German Parliament,
what may be termed the *Verfassungs-bau-versammlung* (there's
a word for you) was to meet at Frankfort last Friday, so to-
morrow we may expect to hear something about it. It is
delightful to think that our dear old friend Arndt, who thirty-
five years ago did so much for the freedom of his *Vaterland,* and
to the eternal disgrace of the late King of Prussia, was perse-
cuted for so many years, is one of the deputies of Bonn, and
he is not unlikely to live to see realized the hope expressed in
his noble song, that the *Vaterland des Deutschen* will be " *so weit
die Deutche Zunge klingt.*" The seven professors who were
deprived of their situations at Göttingen, by that type of
despots, our Duke of Cumberland, are still living, and four of
them are deputies to the assembly. It is expected that the
King of Prussia will be chosen as Emperor of Germany. We
all look with anxious hopes to the liberation of Italy from the
Austrian rule, and as yet all seems promising, although the
struggle may be long and severe. What Government will be
finally established in Lombardy, no one can venture to predict.
I fear too, that in other parts of Italy, especially in the Roman
States, the representative Governments will work very badly
at first; one must hope not so badly as to induce the people
to take back their old despotisms.

You ask me what is doing in science this winter. I can
only speak of the Geological Society, which is going on very
steadily and well. Several good papers have been read, but

none very remarkable in point of interest. The most so was a communication from Mantell, accompanying a very fine collection of bones of the Dinornis sent from New Zealand by his son, in very perfect preservation, and making out nearly the entire skeleton. Another great discovery is to be communicated by Mantell next meeting, the lower jaw of the Iguanodon recently dug up. You are probably aware that a fossil reptile of such enormous dimensions was made out by fragments of thigh bones, and claws, and that no part of the head had been found. The recent discovery is in harmony with these detached portions, and shews that the jaw must have belonged to an animal as gigantic as the dimensions that have been assigned to him.

You will soon part company with our dearest Susan, and I fervently pray, that she may have a safe and prosperous voyage. I am sure you will both regret the loss of so cheerful, so sensible, so intelligent a companion.

The Harley Street Lyells are very well, he looked, I hear, very well at the Levee in his Deputy Lieutenant's uniform. I did not see him, but I saw Mary dressed for the Court last week, and she became her dress well. They appear to have enjoyed the ball at Buckingham Palace last Friday very much.

[In August Mr. Horner attended the British Association at Swansea.]

To his Wife.

Singleton, near Swansea, *9th August*, 1848.

MY DEAREST ANNE,—I had a very pleasant day's journey from Hatfield to Cardiff, by Ross, Monmouth, Usk and Newport. I luckily found a place in the mail at Cardiff and I had scarcely got down from the coach when I saw in the

street Mr. Dillwyn, a friend of the earliest days of the
Geological Society, and we renewed our acquaintance. He
took me under his care immediately, and we went to the
reception-room, the place of information; very few had
arrived, but in half an hour Grove made his appearance, just
come from the circuit at Brecon. Mr. Dillwyn took me and
my luggage in his carriage to Mr. Vivian's, about two miles
from the town, where I met with a most hearty welcome from
Mr. and Mrs. Vivian, the latter shewing me to my room. It
is a beautiful, I may say, splendid house, on a rising ground,
commanding a view of the bay and the opposite coast of
Devonshire. It is modern Gothic or castellated, fitted up
with great comfort, and very handsomely, with thousands of
objects of antiquity and art all over it, and some good
pictures.

But what I most admire is a beautiful conservatory and the
extensive flower gardens round the house. Mrs Vivian is a
very pleasing person, and he is a hearty, sensible, well-
informed man. They have a large family.

There is to be a very large party in the house; already
there are Lord Northampton, Sir Robert Inglis, Lord Adare,
Greenough, Mr. Harford of Blaze Castle near Bristol,
R. Hutton, John Taylor and his son, Lady Ashley and her
daughter. We did not sit down to dinner till a quarter before
nine; we had a splendid feast, the class of which may be
typefied by the great dishes of turtle and venison.

I had some pleasant talk with Lady Ashley, who is very
handsome, and we had much good conversation after dinner,
and in the evening. Sir Robert Inglis was very agreeable. I
was not in bed till half-past twelve, and have been up since
half-past five, for I got, yesterday, a letter from the Home
Office which must be answered soon, but with great care, and
at considerable length.

The weather is very unsettled, much rain—at present there

is sunshine. My leg is tolerably well, not by any means in good geological order.

Ever, my dearest Anne,

Affectionately yours,

LEONARD HORNER.

To his Wife.

Singleton, 11*th August*, 1848.

MY DEAREST ANNE,—I was unable to do more than send you a few lines yesterday. I got my long dispatch to Sir George Grey finished. I must now take up my account from Wednesday morning before breakfast, at which time my first British Association bulletin broke off. But first generally as to this house, where there has been a considerable increase of visitors: Sir John and Lady Charlotte Guest, Sir Charles Lemon, Sir Thomas Acland, Sir David Brewster, Faraday. We sit down twenty-six to breakfast, of whom twenty are strangers, so that you see Mr. and Mrs. Vivian have been most hospitable, and everything is most comfortably arranged in all respects, and his carriages convey us at different times of the day to and from Swansea. Considering the locality, the attendance is very good; I have not yet seen any printed list. Among the foreigners are our old friend Professor Plücker of Bonn, Dr. Forchhammer of Kiel, and Mr. Meyer, Prince Albert's librarian. Yesterday came also the Viennese travellers (who brought me a letter from Heidinger) Ritter, and Von Hauer, also Dr. Morig Hôrnes, and Henry Rogers, who is hardly to be considered a foreigner. He tells me he has been much pleased with his tour in Scotland. Many regrets are expressed that Charles Lyell and Mary are not here.

Before dinner Mr. Vivian took Lord Northampton, Sir Robert Inglis and myself to visit about four miles off at Oystermouth, a fine ruined castle belonging to the Duke of

Beaufort. He is rich in such things, in this part of the
country, for Chepstow Castle, Raglan Castle, and Tintern
Abbey belong to him.

Yesterday, after a pleasant, sociable breakfast, we went at
ten, each to attend his separate Sectional Committee. I went
first to the Section of Chemistry, where there was a communication by Mr. Hunt, of the Museum of Practical Geology, on
the growth of ferns in air, with an increased proportion of
carbonic acid, with reference to the luxuriant growth of that
class of plants in the coal formations. The communication
itself was interesting, and there was a good discussion, in
which Faraday and Lyon Playfair took part. I then went to
the Geological Section, the chief subject being the South
Wales coal field. After some talk in the reception room, I
went with Mr. Dillwyn, Brewster and Henry Rogers to
Singleton, dressed and returned with Sir Robert Inglis, Lord
Adare, and Greenough to dine at the Ordinary Lord
Northampton presiding, where I sat between Mr. Meyer and
Professor Forchhammer. I was victimized, for on account of
Brewster not being there, Lord Northampton coupled my
name with the toast of the University of Edinburgh. In the
evening there was a discourse on the metallurgic process
carried on at Swansea by Dr. Percy of Birmingham, who had
come here ten days before the meeting, to get up this paper.

My dearest Anne,

Your affectionate husband,

LEONARD HORNER.

———

To his Wife.

Singleton, *August* 13*th*, 1848.

On Friday, after our pleasant breakfast, we went into the
town, Mr. Vivian driving me in his pony carriage, a drizzling
wet day, most disappointing, for in the afternoon the beautiful

grounds and gardens of Singleton were to be thrown open. I
went at once to the Geological Section, and there were some
curious details given as to the produce of this coal field. It
was stated that the exports amount to 1,700,000 tons
annually, the consumption of 159 iron and copper works,
about 2,000,000, and that for domestic and other purposes
about 750,000 tons, together about 4,500,000 annually, and
yet such is the extent and richness of the field, that taking at
the most moderate estimation, that vast consumption may be
continued for 1,500 years without exhausting it. Thus we
may feel at ease for our posterity, and when this source is
dried up, they have an inexhaustible Ohio coal basin to
resort to.

Yesterday was devoted to the excursions and the day was
fine and warm, but not much bright sunshine. I chose a visit to
Pentlergare, the seat of Mr. Llewellyn, the eldest son of my
friend Mr. Dillwyn, who got the estate at twenty-one by right
of his mother It is about five miles from Swansea. Mr.
Vivian placed his pony-carriage at the disposal of Sir David
Brewster and myself, and when we got into Swansea, Lord
Adare joined us, occupying the little seat behind. I enacted
charioteer, and my arms ached at the end of the day, for the
pony's mouth was hard as the rock of Gibraltar, and having a
tendency to run away, I had to keep a tight hand upon him.
Mr. Llewellyn is a good chemist, physicist and botanist. A
large party collected at the house, which is situated in a
beautiful wooded dell. The grounds are well laid out, kept in
high order, and the evergreens and flowers in luxuriant growth.
The great point of scientific attraction was a boat upon a lake
which is set in motion by a screw at the stern, worked by a
galvanic battery, an experiment first suggested by Jacobi of
Petersburg. I had not a sail, nor a row, but a propulsion in
the boat round the lake. It is a pretty experiment, but can
hardly be ever applied practicably, for several reasons.

Wheatstone exhibited his clock, in which the hours of the day are ascertained by polarized light, and he had also his wonderful and beautiful instrument, the stereoscope, by which drawings are shewn in relief, thus, a photograph of Danneker's "Ariadne" is made to stand out so perfectly in relief, that you could not distinguish the drawing from the figure itself. We got home at five and I had to dress to be at Mr. Dillwyn's, a mile and half off, by half past five—quick work.

God bless you all.

Most affectionately yours,

LEONARD HORNER.

To his Wife.

Singleton, *August 13th,* 1848.

MY DEAREST ANNE,—I left off this morning, at my dinner yesterday at Mr. Dillwyn's. He has a very pretty house and grounds, beautiful evergreens of enormous size, the arbutus twenty feet high. He has some good pictures too, a charming Murillo. From his house we went to the conversazione, which was held in the new National School, very nicely arranged for the occasion, I had much talk with many friends: Dr. and Mrs. Fowler, Colonel Sabine, Mr. and Mrs. Grove, Buckland and many others. Sir John and Lady Charlotte Guest left yesterday, and Dr. and Mrs. Whewell and Richard Milnes have arrived since I last gave you the names of the guests. Faraday has gone back to London. This morning, after the usual agreeable breakfast, a large party went to church. The Bishop of St. David's preached; the matter was very good, on the advantages of education and the cultivation of science. I came back immediately after church in the dog-cart, driving it, young Ashley and one of the Vivian boys, my companions; the latter very indignant that I drove so slow as to allow a jaunting car to pass us; he is a fine merry boy.

The Dean of Ely and his wife, and Buckland are to dine here. I have seen a great deal of Brewster, for he is in the next room to me, and have found him most friendly and agreeable; I have had much pleasure in renewing our old intimacy. I like Lord Adare very much, I have had a great deal of conversation with him about Ireland, about which he has very sound and temperate views. His father, the Earl of Dunraven, resides constantly in the County of Limerick. He gives a sad picture of the country's poverty, and of the enormous difficulties in doing any good.

I am meeting with kindness from everyone, and am passing my time most agreeably. I have but one regret, viz., that you and one of my dear girls are not with me to share the pleasure, I cannot discard the feeling of selfishness of taking so much enjoyment alone. Sir John and Lady Charlotte Guest pressed me so kindly to visit them at Dowlais, near their great iron works of Merthyr Tydvil, that I could not resist, especially as there will be a very nice party. On Wednesday morning we leave this, and Brewster, Sabine, Wheatstone, and Henry Rogers and I go together, and as Lady Charlotte Guest asked Brewster and me to bring as many with us as we should like to be of the party, we are going to take Plücker, Sijistrom from Sweden, Forchhammer, Von Hauer, and Dr. Hörnes, the two last are the Viennese recommended to me by Heidinger. We are to hire *an omnibus*. The distance is twenty-five miles. I shall leave on Friday morning, by the railroad to Cardiff and thence to Bristol; it is possible I may get home that evening. The dinner party yesterday included a young Prince Metternich, son of the celebrated minister, Baron Hügel, also an Austrian, Colonel Sykes and Richard Taylor the printer, brother of John Taylor who is in the house.

Believe me, my very dear wife,

Your affectionate

Leonard Horner.

To his Wife.

Dowlais House, Merthyr Tydvil, *August 17th,* 1848.

MY DEAREST ANNE,—My last letter left off when I was going to dine at the "Red Lion." It is a club formed at Birmingham by some members of the Association of the more active naturalists, Forbes being a leading one among them. I had often heard of their merriment, "High Jinks," and Oldham of Dublin having given me an invitation, I accepted. Thirty-six sat down to dinner, and a more joyous *Vive-la bagatelle* company could not well be. Monckton Milnes was of the party, and gave us a most funny song, descriptive of the dislike to each other of the inhabitants on the two sides of the Mississippi, with a *nasal* accompaniment. There was much smoking, so that we were at one time enveloped in a cloud. Milnes and I went out to Singleton, between ten and eleven. We found a party (who had been to the evening meeting where Carpenter gave a lecture on some results of the microscope) at tea, and Milnes had scarcely sat down next Lady Ashley, when she exclaimed, "You have been smoking!" and got up and left the room. If every lady would do the same when any gentleman comes near them who has committed the offence, that barbarian abomination would probably be got rid of.

On Tuesday morning, after some preparation for a short introduction to the reading of Charles Bunbury's letter, I went to the Geological Section. The first communication was one from Professor Rogers, a general view of the geology of the United States, confining himself, of course, to the great features, which he illustrated with a very large series of admirable illustrations. He spoke nearly two hours, in a calm, continuous flow of the most perspicuous, graphic language. He was listened to with the most profound attention. I do not hesitate to say that it was the most interesting communication to that Section, during the whole meeting. There was afterwards the Committee of recom-

mendations, and then just time left to go to Singleton to dress
and get back for the Mayor's dinner. The Civic Chair of
Swansea is at present filled by De la Beche's son-in-law, Mr.
Llewellyn Dillwyn, a very gentlemanlike, pleasant young man.
The people of Swansea had subscribed nine hundred pounds
to receive the Association properly, but besides this the
Corporation placed five hundred pounds at the disposal of the
Mayor, and he invited all the strangers to a very handsome
dinner. We must have sat down not less than one hundred
and fifty, and turtle, venison, champagne, and everything else
in keeping, were in abundance. The Mayor was supported
by Lord Northampton and Lord Adare, the latter being a
native of Glamorganshire and one of the members for the
county. By-the-way, I may here add, that I formed a very
high opinion of him, sensible, accomplished, unpretending,
gentle and kind, he must win good opinions everywhere.
After the dinner there was a soirée, and I suppose every one
was there. The chief attractions, were some very fine micro-
scopes, the wonders of which were exhibited by Carpenter and
others. Yesterday we met a diminished party, several having
departed. It rained in torrents at seven, but cleared up for a
time, but there was more rain than fair weather all day.
After a short time in the Geological Section, where Buckland
was discoursing on the Glaciers of Snowdon, the party for Sir
John Guest's assembled at the reception room, where an
omnibus was drawn up. We started at one, some, in some
out; Brewster, Wheatstone, Roget, Rogers, Binney, Warrington,
Plücker, Siljistrom of Stockholm, Von Hauer, Hörnes, and
myself. We were very merry inside, and *conundrums* were
the order of the day, Brewster giving us many most capital
ones, he was very merry. I had some very agreeable and
instructive geological talk with Rogers. Binney is a young
man who is travelling with him, Amos Binney, son of a
naturalist of that name, lately dead, who had prepared

materials for a great work on the land Mollusea of North
America, which is now getting up at Paris, and his son is
going there to look after it.

Warrington is an analytical chemist to the company of
Apothecaries, highly esteemed; he was formerly assistant to
Dr. Turner at the London University, and I have known him
since that time. The rest you know.

The valley we ascended from Neath for about ten miles, is
very pretty, but as we got to the higher part, the country
became wild and bleak all the way to this house. We did not
arrive till past seven. Sir John and Lady Charlotte Guest
met us most heartily, and they located their eleven guests in
their respective rooms, which took some time. We sat down
about half-past eight to a most sumptuous dinner, in a style
in no degree inferior to what I have seen at Holland House,
and Lansdowne House, not excepting our dining off plate.
The house stands alone by the works: there are a dozen flaming
furnaces within gun-shot. Several of us went out at twelve
o'clock to look at the wonderful sight. It had the appearance
of a great town in flames in some places, in others like the
workshops of Vulcan with the Cyclops forging and hammering
on the most gigantic scale. I sat next Lady Charlotte Guest
at dinner, and found her very agreeable. It is now time for
me to go down to breakfast.

On my arrival I found your dear letter of Tuesday, which
keeps my mind at ease about my treasures at Rivermede.

It is not impossible that you may see Forbes at dinner on
Saturday, and stay all night. God bless you, my very dear
Anne, and with love to my dear ones with you.

> I am ever,
>   Your affectionate husband,
>   LEONARD HORNER.

To his Wife.

Dowlais House, *August* 18*th*, 1848.

MY DEAREST ANNE,—I wrote to you yesterday morning. After a pleasant breakfast, when I sat between Mr. Rogers and his friend, Mr. Binney, and after some sauntering (the day being very wet, we hoped by waiting that it would clear up), we set out to see the works, or I should rather say a part, and were shown them by a most intelligent manager. The steam engine which works the bellows, which send forth blasts like Æolus (and his breath is sometimes very hot) into the fiery furnaces, is the largest I ever saw; it has a power of three hundred horses, and the blast cylinder, that great body of the bellows where the propelling force is applied, is twelve feet in diameter. Still this is not the largest engine in the world; that now at work to drain the Haerlem See is more powerful: it was made by the same man who erected Sir John Guest's, I forget his name, but he lives at Hoyle in Cornwall, and there the engines were made.

After seeing the works we returned to luncheon, and then set off in a sort of covered waggon, drawn on a railroad that goes to some lime quarries, by a horse. Our party was Lady Charlotte Guest, Buckland and Daubeny (who had arrived a quarter of an hour before), Roget, Wheatstone, Brewster, Rogers and myself. Others went some other way. The quarries are very large and are worked solely for smelting the iron. We did not get home till past seven, so there was a hurry-scurry to dress, as some additional guests had been invited to dinner at seven. My neighbours were a Mr. Hutchins, a nephew of Sir John Guest, a lively intelligent man, and Mr. Layard, the Eastern traveller, who has made so many wonderful discoveries at Nineveh, and who was very entertaining. In the evening Lady Charlotte Guest brought forward a harper, three men and a woman, belonging to their works, and we had some charming Welsh music; the woman

had a very fine voice. Then we had two young men who
danced before us Welsh Jigs, most capitally. These amuse-
ments were followed by the younger members of the family
dancing to the harp. It was near one before I got to bed.

Sir John and Lady Charlotte Guest have ten children, seven
at home just now, and most of them appeared, together with
the tutor, the English governess, and the French governess;
so that altogether there is a pretty considerable party. I
reckoned up thirty-three, besides servants, who must have
slept in the house last night. Before going to bed, a party of
us went out to see the blazing fires around us, as we had done
the night before, and a splendid sight it was.

We have had a very merry breakfast, I was at Lady
Charlotte Guest's end of the table, and we had Buckland,
Brewster and Rogers, all talking very cleverly and agreeably,
Lady Charlotte Guest being behind neither of them in spirit
and humour. This forenoon Rogers is going to give a lecture
in Lady Charlotte Guest's school, where we are all going.

The wind has changed to the East, and we have a clear
sky.

After my letter of yesterday was sealed, I got yours, my
dearest love, and Nora's, which made the day brighter to me.
I trust that I am now very soon to see you all, and truly
happy shall I be to get to Rivermede.

Your ever affectionate,
LEONARD HORNER.

To Sir Charles J. F. Bunbury, Esq.

Rivermede, *August* 28*th*, 1848.

MY DEAR CHARLES,—I have to thank you and my dear
Frances for letters which I received very lately from you and
her. It was a great comfort to me that you had arrived on
this side of the Alps, for although you had never been near

the seat of war, there must be a good deal of disorganisation, and travelling must consequently be less secure. Your and her letters are most interesting, and you have both been very good in keeping us so fully informed of all your proceedings. I can enter well into your disappointment at the sad turn which political events in Italy have taken, after the cheering prospect that a brighter day was dawning. There was wanting, I fear, a right feeling in the great *mass* of the people, for had they been firm and resolved, even the disciplined army of Austria must have given way. But perhaps Austria has been made to feel how little progress she has made in the last thirty years in gaining the affections of the people, that consequently her reserve is weak, and that she may take the opportunity of an honourable and quiet abandonment of a great part of the country she has hitherto held. You will probably hear much of Swiss affairs, how far that country is settling into a quiet, peaceful state, whether the movements there a year or two ago have been productive of any advantage to the people likely to be lasting.

I will try to give you a short sketch of the Swansea Meeting. I left Manchester on the 3rd and passed that evening and the next day at my favourite place, Malvern. I offered to pay a visit to my friends the Henry's at Haffield near Ledbury, and they arrived from Aberystwith the day I went to them. I passed my time very agreeably, they have travelled much, particularly in Italy, and they have a fine library and a large collection of valuable prints. Their house is situated close by Eastnor Park, and you know what a charming country that is. I got into a Worcester coach at Ledbury for Cardiff, and next day got to Swansea by the mail. My friend Grove procured me an invitation from Mr. Vivian, the father of Mrs. William Gibson Craig, and after calling at the reception room I met my old friend, Lewis Weston Dillwyn, whose name as a botanist you are perhaps

acquainted with. He was one of the first list of Honorary
Members of the Geological Society. The meeting has been
considered eminently successful, and more agreeable than
many previous ones. In the Geological Section, the best
communication we had was from Professor Henry Rogers
of Boston, who gave a general view of the structure
of the United States, touching chiefly upon the great
features (so well described by Lyell), the elevations and
flexures of the Apallachian Chain, the great drift, and the
vast coal fields. He spoke nearly two hours, and had the
most ample and admirable illustrations to refer to. I never
heard a more clear delivery, every word was well chosen, and
all his descriptions were most graphic. We had good papers
from Forbes, Ramsay, and Oldham. Your letter to me,
describing the coal plants of the Tarentaise, reached me on
the 13th, and I read that description to the Section on the
15th, introducing it by reading an extract from my address of
1846, where I spoke of that anomalous occurrence of coal
plants and belemnites. Your account of your examination
excited a good deal of interest. Saturday, the 12th, was
devoted to excursions and fortunately it was fair. I went
with Brewster, Wheatstone, Lord Adare and some others to
Pentlergare, a place belonging to a son of Mr. Dillwyn, a good
chemist, physicist and botanist. It is a beautiful place, in a
wooded dell, with extensive, well kept grounds and gardens.
Plücker has made some interesting observations on the
dia-magnetism of certain minerals, which Faraday told
me he considered very important as illustrative of that principle. The next meeting is to be at Birmingham, in September,
1849 ; Dr. Robinson of Armagh, President. I hope that you
and Frances will be there. I accepted an invitation from Sir
John and Lady Charlotte Guest, to visit them at Dowlais
House, near Merthyr Tydvil. You are aware that his Iron
Works there are the most extensive in England. It will give

you some idea of their extent when I tell you that he consumes above 1,400 tons of coals *daily*. When visiting the vast furnaces, some of which have been in action without intermission for fourteen years, it occurred to me that some very valuable experiments might be made on larger masses, and for a length of time of continued heat far beyond anything possible in our laboratories, with the view of illustrating the theory of metamorphic action. Sir John willingly consented to my proposal, and Rogers, Daubeny, Warrington and I, laid our heads together, and drew up some notes of subjects for investigation, and we have now to consider the best mode of conducting the experiments. I am going to write to Dr. Percy of Birmingham on the subject, whose advice will be of great use.

I have not heard lately of any news of Joseph Hooker, but the last was very good : he was proceeding to the Himalayas. A second volume of the " Memoirs of the Geological Survey " is just out. It contains a paper by Hooker on coal plants which from the glance I have had of it, appears to be very interesting and valuable. It is admirably illustrated.

W. R. Grove, Esq.

Rivermede, Hampton Wick, *September 12th,* 1848.

MY DEAR GROVE,—You may remember my having told you at Swansea that I had spoken to two men at the meeting about becoming candidates for the R. S. One of them was Captain Ibbetson, but the other I cannot remember. I wrote lately to Dr. Andrews of Belfast on the same subject, and he is well pleased to be proposed.

I shall be very anxious until I hear what names have been selected for the next Council, and that there is a majority in the present Council favourable to the selection of men who will carry out the recent reforms in a good spirit, and will assist

in rendering the meetings more attractive, and be otherwise
active in promoting and trying to raise the character of the
Society.

I look back with unmixed pleasure to the meeting at
Swansea, and I know how much we are all indebted to you
for the success of it. I have heard that the dinner given in
your honour went off exceedingly well. I trust that, now all
is past, you have not a single feeling connected with the
meeting that is otherwise than entirely satisfactory, and that
you reflect with pleasure on all the pains you bestowed and
the time you devoted to get it up. You may think that I
view with a partial eye what you do, but I assure you that I
have said no more than what I have heard expressed by every
one who was at Swansea, with whom I have conversed on the
subject. The two days at Dowlais House were very agreeable.

On Saturday I set out for Lancashire on my autumn tour
of inspection. I shall not get back for the first meeting of
the Philosophic Club, but I will do all I can to return in
time for the second meeting. Mrs. Horner and one of my
daughters accompany me.

I hope Mrs. Grove and all your children are well. Somehow
or other there has been great mismanagement on our part,
that we have not had Mrs. Grove and the children here, for I
am sure she would have not refused our invitation. I can
only account for it by the small number of summer days, and
our mutual absences from home.

I am, my dear Grove, faithfully yours,

LEONARD HORNER.

To Charles Lyell, Esq.
Manchester, 23*rd September*, 1848.

MY DEAR CHARLES,—I received this morning a letter from
Mr. Burton, Advocate, Edinburgh, who lately published a life
of Hume, that he had undertaken at Chambers' request to

prepare the abridged edition of my brother's Memoirs, and I
believe the task could not be in better hands. This puts me
in mind to tell you of a speech made to me on Thursday by
Mr. Vernon Harcourt's youngest son. We were alone, and as he
is at present an undergraduate at Trinity College, Cambridge,
I began to talk of the lectures there, and asked if he attended
Sedgwick's course. He said that he had, and then spoke with
regret of the want of encouragement given to Natural Science,
and the neglect of the professional method of instruction.
After some further observations, he added, " But it is to you,
Mr. Horner, that I am indebted for the opinions on this
subject which I hold," and on my expressing my inability to
comprehend how that could be, he added, "It is your
memoirs of your brother which have impressed me with the
conviction how defective the system of our Universities is
for the business of life." This, you may well suppose, was
very gratifying to me to hear. It is an effect produced by
that work which I had never before heard of, and it may not
be the only instance. The family spent last winter and spring
in Madeira, and Harcourt has brought home some very
curious things from the island, illustrative of its geological
features, but his most interesting specimens are from the
small island of Porto Santo, thirty miles distant, a limestone
full of fossils, and I suspect of comparatively modern age,
perhaps miocene, converted into perfect Carrara marble.
Conceive corals made of white sugar candy, in minute crystals,
and you will have some idea of the thing. It would be worth
your while to return by York to see these collections as well
as specimens from Teneriffe, which he also visited. There is
much to interest you in the Museum, which Charlesworth,
whom I saw, appeared to be putting in excellent order. You
are a great economist of time and opportunities, so why
spend two days on the trackless waste of the sea? What you
would see at York would be well worth the difference of

expense. Susan Lloyd would be delighted to lodge you
during your stay, and the Vernon Harcourts are to be in
York till February. They are most agreeable people.

To his Daughter.

Bury, *October* 26*th,* 1848.

MY DEAREST KATHARINE,—I have been an unkind father not
to have written to you for so long a time. I have rested too
much on my belief that you knew me to love you so tenderly
and warmly, that you could not think my silence any
indication that you did not occupy my thoughts a great deal.
As usual, I brought a good supply of books, and according to
my usual experience, have found no time to read any of them.
Besides my ordinary occupation when in circuit, and the
necessity of writing a great deal in the evening, to prevent
that, which I of all things dislike, *viz.*, an accumulation of
arrears, I voluntarily undertook an inquiry which I
considered of importance to make, in order to ascertain the
feelings of the workpeople of the factories as to the ten hour
Act, whether the law passed ostensibly for their benefit is
considered to be so by themselves; I have conversed with two
hundred and sixty-three individuals, and taken notes of their
evidence, and the writing it out for the press has occupied a
great deal of time. I cannot yet speak with any accuracy,
but my general impression, from my own enquiries, and from
those of the five Sub-Inspectors of my district, is, that the
work-people prefer the shorter time, although they get less
wages, in a majority of the cases. It is a result I did not
expect, and is a justification of Lord Ashley, and of those who
urged him to persevere, that I was not disposed to give them
credit for. I never for a moment doubted the purity of Lord
Ashley's motives, but I thought that he had been misled to a
great extent, by designing people. I feel quite sure of this,
that there never will be a return to twelve hours of work.

The time is now approaching when we shall move homeward, and Rivermede begins to dawn after a long night. But
that night has passed very pleasantly, and in something
more substantial than dreams; very agreeable realities of
which your Mamma and Joanna have been the life. The
difference is not to be told between the solitude of my
inspecting hours when I come by myself, and the cheerful
faces that greet me when I come home in the evening. With
my best love to Harry, and kindest regards to Mr. and Mrs.
Lyell and the whole sisterhood.

I am, my dearest Katharine,

Your affectionate father,

LEONARD HORNER.

To his Daughter.

Manchester, *November* 7th, 1848.

MY DEAREST KATHARINE,—We set out to-morrow at midday
for Oulton Park; on Friday we go to Dudmaston, and on
Monday evening I trust we shall drink tea in Harley Street.
We shall thus arrive to celebrate Charles's birthday the next
day.

Although I have been very busy, I have, thank God, been
very well all the time I have been in Manchester; I have not
been prevented by illness a single day from going out, and
have been very little troubled by my weak legs, indeed not at
all for the last month. You know that there is perfect
sunshine when your dear Mamma is near me, and in Joanna
I have had a bright star too, rendering even the sunlight more
cheerful. We have had some occasional society that has
afforded us interest and amusement, and our new acquaintances,
Mr. and Mrs. Schwabe, we find very agreeable and kind.

We have just come home from spending a very pleasant
evening with Dr. Mainzer, who gave us some exquisite music.

I find him an extremely well-informed, able, independent-minded man, and am altogether much pleased with him.

I wish you joy, with all my heart, of your botanical dis-covery.* Consult Mr. Lyell whether you ought not to send an account of it to Sir William Hooker, for his Botanical Magazine. This event will encourage you in your study of this very agreeable branch of natural history, which I doubt not will prove an endless source of agreeable occupation to you in India.

I am very sorry that you give so indifferent an account of Mr. Lyell, and trust that when this reaches you, he will be better, and not enter upon winter with a cold. We have really winter weather here; I went about twelve miles towards the Derbyshire hills this morning, and the ground was powdered with snow.

My best love to Harry, and kindest regards to Mr. and Mrs. Lyell.

Your affectionate father,
LEONARD HORNER.

---

* Of a rare moss.

CHAPTER VII.

1849—1850.

To his Daughter.

Rivermede, 16*th June*, 1849.

My Dearest Frances,—Unless I am kept in town by Sir George Grey, I mean to set out for Lancashire on the 22nd.

I think the drives in our phaeton have done me good, they are at all events very enjoyable, and they become more so, the better we become acquainted with the beautiful country around us. We had a very successful day at Box Hill on the 11th, *barring* the east wind. I never knew a more ungenial season, and there is scarcely a plant in our garden that is not more or less blighted, and the gardener says he never in his life swept dead leaves off a lawn at midsummer before this year. Still the country is in great beauty and all the corn crops look well ; and if rain keeps off, there will be an abundant hay harvest all round London.

We had Mrs. Jameson with us yesterday, and we took her to Lady Byron's at Esher to-day. She (Mrs. Jameson) was very agreeable. Her conversation was excellent, lively and clever; her opinions are good on all subjects, and she is often very entertaining. She passed six months in Ireland last year, and she gives a sad picture of the widely spread misery.

What horrible things are going on in so many parts of Europe ; the worst of all in my opinion, is the disgraceful and unjustifiable attack on Rome by the French, which ought to bring down upon them the indignation of the civilized world

Mrs. Schwabe, who has lately come from Germany, gives a sad account of the state of the country, and of the military depotism now reigning in Prussia and Saxony, where the soldiery have committed the most horrible excesses, especially in Dresden and Leipzig.

I am looking forward with much pleasure to a visit to Mildenhall this summer.

God bless you both,
Your affectionate father,
LEONARD HORNER.

To his Wife.

Bonaly, *July 15th*, 1849.

Here I am in this most beautiful spot, with our old kind friends, and old associations of many happy days.

I wrote to you on Thursday night from Greta Bank. My visit there was very agreeable ; they are both pleasant, and he* particularly sensible and companionable; and then the place and country round are so charming. They begged me to bring you soon, and said that they can take us all in. The first day they were alone. On Friday a Rev. Mr. Martineau a relation of Harriet, and his wife, sister of Smith O'Brien, dined there, and another clergyman, a pastor in some neighbouring village. The conversation was good, and fairly liberal. I left yesterday morning by a coach to Penrith at ten, waited a full hour there for a train, and then got on by Carlisle and the Caledonian railway, arriving at Slateford about a quarter before eight. The day had been very hot, but as we approached Edinburgh, there poured up from the sea, driven by an east wind, a dense cold fog, which obscured every thing; I found the drosky waiting and "the immortal Geordie" turned out to be *the horse*, which has drawn that drosky for 23 years. The driver, an oldish man, Philip, is a capital specimen of a genuine Scotch peasant. We had much talk all the way, the crops and Free Kirk being the chief topics,

°Mr. Thomas Spedding.

his remarks were very shrewd and his dialect, pure unadul-
terated Scotch. "Look, Sir, what a grand crop o' tawties."
Cockburn and George were dining at a Mr. Marshall's near
Currie, and did not return till eleven. In the meantime I had
a long talk with Mrs. Cockburn about many old friends.
Macbean was here at tea, three or four days ago, very cheerful,
but not much improved ; Jeffrey has been very ill, is better,
but still very weak ; Mrs. Jeffrey nearly well, and walks about
with ease. This morning I walked round the place, going to
the top of *Pisgah.* I am quite lost in admiration of the growth
of the evergreens and trees. I do not think that in the most
favoured spots of the South of England, the evergreens could
grow more luxuriously. The effect now that the plantations
and shrubs are in full growth, shews the skill and taste with
which the plan was conceived and executed. I do not wonder
at his grudging every hour he is absent from so charming a
spot. Pillans came to breakfast, and we shall have a ramble
on the hills by-and-bye, I make no doubt.

To-morrow I go to Kinnordy, and mean to return to
Edinburgh on Thursday, I have offered to dine with the
Murrays that day, they go next day to Strachur. On Friday
the Cockburns dine with the Thomas Thomsons, and they
have sent a message for me to come too. The Academy
examinations are to be on the 23rd and 24th. The Exhibition
Day will be on the 25th and the High School Examination
will be on the 27th, which I also mean to go to.

Tell dear Susan that Mr. Spedding, who is an intimate
friend of Mr. Laurence, said that Laurence has been for
years working upon the subject of the colours and ground used
by Titian and the Venetian school, and that he thinks he has
made an important discovery. Mr. Spedding said that he
believed Susan would have great pleasure in conversing with
Laurence on the subject ; that he would recommend her to ask
him to meet her at the Exhibition, go over the principal

pictures there with him, and then go to the National Gallery,
and contrast the styles, this he did with Mr. Spedding, and
his lecture was most interesting and instructive.   As Mr.
Laurence lives so near Harley Street, it will be a good opportunity for Susan having some conversations with him.

To his sister, Mrs. Byrne.

Mildenhall, *6ih August*, 1849.

MY DEAREST FANNY,—I will give you *chapter second* of my
journey in Scotland, as I learn that *chapter first* amused you
and Nancy.   I left off at the 23rd of July ; that day the
Academy Examinations began, and the Exhibition Day was
on Wednesday.   Nothing can be going on better.   The new
rector, Mr. Hannah from Oxford, is a capital man.   On the
Wednesday morning I was at a breakfast given by the old
pupils of the Academy who are associated as the Academical
Club, and the same day there was a dinner of the Directors
and Masters, and some of the parents of the duxes.   The Lord
President Boyle's son, was the dux of the whole school.

Among my old associates in that work, I met, besides
Cockburn, Lord Moncrieff, John Russell and Richard Mackenzie.
I was most heartily greeted, and after dinner my health was
proposed in a very kind manner by Dr. Tait of Rugby, whom
I took in a blue jacket and white trousers to the Exhibition
Day, when, as Archy Tait, he was dux of the whole school in
1826.   I dined on the 23rd with Solicitor General Maitland,
and on Tuesday with Pillans in his new house in Inverleith
Row.   We had Thomas Thomson, and among several others,
the worthy Bryson, the watchmaker, my right hand man in
establishing the School of Arts.   My friend Nightingale was
also there, and Mrs. Nightingale and her daughter Florence
came in the evening.

Between the termination of the Exhibition at the Academy,
and the dinner on Wednesday, Cockburn and I went to

Craigcrook. I thought Jeffrey much changed, pale, emaciated and feeble in body ; but his mind is in perfect preservation. He has had a severe illness, and for the first time while I was there, walked out in the garden among his roses, and I was glad to see that his step was firm. He had been reading Lyell's new book,* and spoke of it with the highest praise. Mrs. Jeffrey is also a good deal aged, but she is very well, and cheerful and happy-tempered as ever. Charlotte, her husband and children were all there. The place is more beautiful than ever, and the show of roses and other flowers was splendid.

On the 26th, Cockburn, Nightingale, Florence and I, set out for the pier of Leith, which is now extended to the Martello Tower very nearly, and a glorious walk it is of more than half a mile from the shore. The day was beautiful, the sea smooth as glass, and you know what a view there is looking back upon Edinburgh. Coming up Leith Walk I started for the south side of the town, where I had not yet been. I walked up College Street, through Lothian Street into Park Place, stopped at the old house, looked at my mother's bedroom, and the kitchen, and thought I heard the cuckoo clock, and peeped over the wall into the garden. I called on James Tait† who still lives in the house inhabited by his father and grandfather ; his family have been in that house more than seventy years, and 1 daresay the brass plate on the door is the oldest of its species in Edinburgh. I know that you and Nancy will be walking along with me, with a vivid recollection of the old places, that stir up many thoughts of pleasure and sorrow.

On Friday I started by the railroad for Glasgow and Greenock. The evening was fine, so that I saw to perfection the view of the Ochills over the Carse of Falkirk, with Ben Lomond clear to the top. I was equally fortunate in seeing the scenery of the Clyde, the railroad coming close to the

* "Second Visit to the United States."
† Brother of the late Archbishop of Canterbury.

river at Dunbarton and continuing so to Greenock. Next
morning I started for Strachur; the morning was hazy, but
it cleared up as we entered Loch Long, and thus I saw the
charming scenery of Ardentinny Bay, the mountains, and
Loch Goil, and the pass by coach from Loch Goil Head to St.
Catherine's on Loch Fyne. There Murray and Robert Graham
met me, and we drove to Strachur, five miles off, where I was
most kindly welcomed by Lady Murray.

Sunday was rainy, but Murray, Graham and I drove seven
miles down Loch Fyne to Castle Lachlaw, which on a fine day
is, I daresay, a pretty place. Monday it poured till two
o'clock, but I occupied myself attending a quarterly examination of a Free Kirk School, which is kept in a barn belonging
to Murray, which he has generously given up as a kirk and
school, although he thinks the *disruption,* as it is called, foolish
and mischievous. I was much pleased to see the school—
about seventy-five—well taught. I examined many of them
myself, and a large proportion of the kilted bare-footed boys
stood a most creditable trial in reading, geography and arith-
metic, and I was happy to see the good old custom of the
Scotch parochial schools kept up, inasmuch as about half-a-
dozen of the little fellows stood a good examination in the
Latin rudiments, which the minister and I put them through.
Next day was fine, and after an early breakfast Murray and I
started for Ardkinglass at the head of Loch Fyne, an exquisite
spot; grand scenery, and a flower garden which could not be
made more perfect anywhere, and few places so well, as there
is a piece of water filled with the pellucid water of a mountain
stream. It belongs to Mr. Callender of Craigforth, who married
the heiress. We then crossed to Inverary and drove seven
miles up Glen Shirra, then through the grounds of Inverary
Castle, where in the garden I saw the famous Portugal laurel,
a dome of foliage twenty feet high, and seventy-five yards in
circumference.

Next day I was obliged to leave, though much pressed to
stay some days longer, nothing could be more kind than
both were. Murray was determined that I should see as much
as possible, so in place of my going back by the coach, he
drove me through the beautiful pass of Glen Eck, and we
called at Glenfinnart, a charming spot. The steamer took me
up at Ardentinny, the mountains of Arran and Ben Lomond
were bright to their summits, and at seven I landed at
Glasgow.

With our united love, I am, my dearest Fanny,

Affectionately yours,

LEONARD HORNER.

To Charles J. F. Bunbury, Esq.

Rivermede, 24th November, 1849

MY DEAR CHARLES,—It gave me much pleasure to receive
your letter of last Sunday, and the good accounts you gave
me of your dear wife are confirmed by your letter to Susan
received this evening. I am delighted to hear too that she is
finding so much pleasure in reading, as that shews strength,
and her remarks in her letter to Nora this morning, on the
character of Cicero, and on the Roman republic, shew that
she is capable of exerting thoughts far beyond light reading.
I hope she will still go on with her project of a translation of the
life of Dante, for there is no pleasure greater than having a
work to carry out, that will occupy some time. It has been a
very great pleasure to her dear mother and Nora, to have had
an opportunity of nursing her during her late severe illness,
and it was a great consolation and comfort to me to be near
her bed so long. We never can tell how strongly we love our
children, until they are in peril. God grant I may never see
her the like again.

I am glad to find that your time passes pleasantly amid the
objects of interest that Edinburgh affords. You speak of

visiting Arthur's Seat, hammer in hand, but you do not mention
having gone there with any geologist who knows the spots of
most interest. If Maclaren is in Edinburgh, I am sure he
would be delighted to pass a long day with you on the hill.
Have you made his acquaintance ? I shall be happy to send
you notes of introduction to him, David Milne, and Hugh
Miller, if you would like it.

I shall be much obliged if in your visits to second-hand
booksellers, you would try to find for me, at a moderate price,
the third edition of Brown's Essay on Inquiry, on Cause and
Effect, which W. Erskine designates as "a magnificent monument of the author's metaphysical genius."

I have read with much interest your father's narrative of
the Campaign in North Holland in 1799, a copy of which he
was good enough to send me. What a picture he draws of the
then state of the British army, and of the rash imbecility of
the Government in sending ont such an expedition under such
leaders ! His sketches of the members of the Council of War
are very graphic. I have had the advantage of reading it with
a map on a large scale, which I brought from Holland in 1814,
where all the small villages he mentions are laid down.

You will be glad to learn that at the Council of the Royal
Society last week, we unanimously voted the Copley medal to
Murchison, for his great services to geology, by his classi-
fication of the older palæozoic rocks. He is much gratified.

My love to my dearest Frances,

Your affectionate father-in-law,

LEONARD HORNER.

To   Charles Bunbury, Esq.

London, 28*th January*, 1850.

MY DEAR CHARLES,—I wrote to my dear Frances yesterday.
Give her my best thanks for so kindly thinking of preventing
too sudden a shock to my feelings, by my hearing accidentally

of the unexpected death of my dear and excellent friend
Jeffrey. All our accounts of him lately had been so cheering,
that I was in hopes he was to be preserved for some time to
come. You will have many opportunities of hearing from
those around you, how much there was to love and venerate,
as well as to admire in his character. It is a blank, that in
the society of which he formed a part, can never be filled up,
and Edinburgh has lost one of her citizens of whom she had
much reason to be proud.

Charles Lyell has allowed me to read your letter to him of
Saturday. I am much interested in all you say on the plants
of the coal formation, and the facts you state are most important elements in marking out the history of those deposits,
one of the most difficult problems which geology has to solve.

In all you say upon the education question I entirely concur.
Most dangerous as I deem it to be, that the secular education
should be at all under the control of the priesthood of any
denomination, I can conceive nothing more dangerous either
to the welfare of society, or to the happiness of the individuals,
than that the children of the great body of the people should
be left without special religious instruction, and I think that
as long as religion is recognised in a country, the great majority
of the country will be of opinion that the Ministers of religion
ought not only not to be excluded, but should be employed to
perform that duty.

I agree, too, in thinking that education ought ever to be a
great department of the Government of a country, in order
that it may be provided to the fullest extent wanted, and of
the best quality, and if the share assigned to the ministers of
religion do not go beyond their special function, I do not fear
injury, but on the contrary augur advantage from their
participation.

Ever affectionately yours,
LEONARD HORNER.

To his Daughter, Mrs. Bunbury.

Manchester, *February 23rd*, 1850.

MY VERY DEAR FRANCES,—Your dear Mamma and Nora arrived safe and well two hours ago. A bright sunshine is spread over and around me. There is a very great degree of excitement among the Factory operatives by the cruel disappointment to their hopes by the decision of the Court of Exchequer. I am inclined to think that it is as well that things have taken the turn they have done, because there must be now an Act of Parliament to settle the question. I do not think that the judgment will much increase the evil of relays, for it had got to a great length, and had the judgment been different, we should have been struggling on with a doubtful, uncertain law upon a point which should be clear and unequivocal. The Government has behaved in a very discreditable way in this matter; so soon as doubt was thrown on the true meaning of the Act, by the contradictory decisions of magistrates (and that was the case twenty months ago) they should have brought the subject before Parliament. If they attempt to infringe upon the Ten Hours' Act, they will be assuredly beaten sooner or later, for it has taken deep root in the good opinion of the operatives, and they will make a great struggle to have the Act honestly and uniformly carried into operation. It quite disgusts me to hear the cold, calculating economists throwing aside all moral considerations, and with entire ignorance of the state of the people who work in factories, talking of its being an infringement of principle to interfere with labour. Why interfere with the use of capital in any way then? and do we not see laws passed every year to check the abuse of the application of capital, when it is productive of great moral and social evils. If I were free to write, I could from my experience make such a statement as would shew the fallacious reasonings, and *bad political* economy, of these very economists, who, with their extravagant extension of their doctrine of *laisser faire*, bring discredit upon the science they cultivate.

I shall not have much time for reading here, but I have brought some geological work with me, even a paper for the Geological Society, so Charles will see that the old soldier is still shouldering his musket. I have brought also the very interesting volume of Sydney Smith's lectures. I think you will both be pleased with them. I read last night his lecture "On the conduct of the understanding," which is very original, and full of practical good sense.

My kindest regards to Lord Cockburn, tell him that I had some conversation with Dr. Charles Bell to-day, about Jeffrey's monument, and that I shall write soon to him on that subject. Mamma and Nora unite in best love to you both with,

My dearest Frances,

Your affectionate father,

Leonard Horner.

———

To his Wife.

Skipton, *2nd May*, 1850.

My duties leave me plenty of time for reading in the evening. I am much interested in our excellent Jeffrey's Essays, I have finished all his reviews of Scott's novels, and Miss Edgeworth's "Tales of Fashionable Life," and they are most interesting and instructive.

I have heard to-day that Sir George Grey has at last had the courage to take the right course in the Factory Bill. There will be a great opposition to his plan among certain classes of the operatives, but as I know that Lord Ashley is, in his heart, favourable to it, there is a strong probability that there will be a decided majority in its support in Parliament. It is by far the best plan that could be adopted ; it secures everything that is valuable for the operatives, it

confers a great additional boon upon them by their being
released from work on Saturdays at two o'clock, and it gives
the Inspectors the most ready and prompt means of detecting
any over-working. It does all this, and it gives the mill
owner two hours more work out of his machinery in the
week, a benefit which the work-people will share in, without
any extra exertion that practically will be felt.

The weather continues fine, and is milder. I hope the
garden is daily improving in beauty.

"My dear wife! thou wast the love of my youth, and have
been much the joy of my life; the most beloved, as well as
most worthy of all my earthly comforts: and the reason of
that love is more thy inward, than thy outward excellencies,
which yet are many, God knows, and thou knowest it, I can
say it was a match of Providence's making. And my dearest,
let me recommend to thy care my dear children, abundantly
beloved of me, as the Lord's blessings, and the sweet pledges
of our mutual and endeared affection."

Thus said William Penn, in the tenth year of his
married life, and so say I in the forty-fourth of mine.

Your devoted husband,
Leonard Horner.

To his Wife.

Liverpool, 12th May, 1850.

I came here on Friday, and after visiting a large mill
yesterday, I called on Melly, and sat some time with him.
This morning I went to Renshaw Street Chapel, and heard a
very good sermon from Mr. Thom, and had some conversation with him afterwards. He is a very agreeable, well-
informed, liberal man. He told me I sat to-day in the very
corner of the pew Blanco White always occupied. When the
sermon was over I found Mrs. Melly waiting to speak to me;
she is in such delicate health, her husband tells me that he is

going to take her to a warm climate next winter, and thinks
that Egypt will be the best place.

This house is full of the passengers by the "Atlantic" that
arrived on Friday evening. I am sorry to say that they are
following the French fashion of wearing beards.

Your affectionate husband,
LEONARD HORNER.

To his Daughter.
Manchester, 17th May, 1850.

MY DEAREST SUSAN,—Do you intend to send a picture to
the Manchester Exhibition, and have you all the information
you are in want of? Can I do anything for you here?
I have heard nothing of "Samuel and Eli" since you got it
back from the Academy.

I came here yesterday, and found everything ready for me,
and very comfortable and I soon settled myself. After
some business matters, I amused myself with a book
Dr. Boott lent me, and indeed almost urged me to read,
—the life of Alexander Bethune, a working man in Fife;
poet, essayist, and political economist. He must have
been a remarkable person, and it is not easy to understand how such poetical feelings, such enlarged and
sound views on the great concerns of life, could have existed
in a hedger and ditcher, whose earnings seem never to have
exceeded seven shillings and sixpence a week, and who never
was at any school or place of education for more than twenty
weeks in his whole life. In these respects he is an object of
wonder. He died in 1843, at the age of thirty-eight. There
is a little volume entitled "Practical Economy explained and
enforced," which I have also from Dr. Boott, written partly by
the same Alexander Bethune, partly by his brother, also a
common labourer. These men are not productions of the Scotch
parochial system, for they seem to have had no instructor

except their mother, but she must have been no ordinary person, from the books she was in the habit of quoting, her favourite author being Cowper.

Ever, my dearest Susan,

Your affectionate father.

LEONARD HORNER.

To Sir Charles Lyell.

Rivermede, August, 1850.

MY DEAR CHARLES,—I thank you for your very interesting geological letter from Berlin of the 8th. I have read the Memoir of Von Buch in our journal since I got your letter and I confess I do not see Von Buch's theory in the light you do. I do not think there is anything improbable in a mass of pasty granite being forced up from below in a dome shape, carrying up with it the incumbent stratified rocks, and shattering them, so that they would be exposed to the action of denuding causes under the sea. Von Buch describes the blocks as being not promiscuously scattered, but arranged in a symetric form, and the ends presenting prominent and re-entering angles, opposite each other. Did you notice this arrangement? These detached blocks, standing so high above the surface, seem to be not very reconcilable with the arrangement as described by Von Buch. What an iron constitution Von Buch must have! a pedestrian tour in his 82nd year!

The discovery of the loess at so great an elevation and at such a distance from the Rhine Valley is most curious and interesting. Altogether you seem to have been not only inde-fatigable in your hunt, but to have run down some capital heads of game. I am glad that you have seen so many good geologists, and that you got to Berlin in time to catch so many of the scientific men before they departed for the enjoyment of their *Ferien*. I shall send your letter on to Bunbury, to whom it will be a great treat at Scarborough. I have no news to give you. The Edinburgh meeting appears to have been eminently successful. There is a report of a very interesting paper by the Duke of Argyll, describing tertiary beds with plants and shells overlaid by vast masses of trap rocks in the Hebrides, so that probably Staffa and the Giant's Causeway are of far more modern date than has generally been supposed. There appears to be no doubt of the accuracy of the facts described, for Smith of Jordanhill examined the spot with the Duke, and Forbes appears to have determined the shells· What period of the tertiary, does not appear in the report I have seen.

We have been extremely interested by dearest Mary's most excellent letter. She makes me quite long to go to Berlin.

God bless you both.

Ever affectionately yours,

LEONARD HORNER.

To Sir Charles Bunbury.

Rivermede, *August 25th*, 1850.

MY DEAR CHARLES,—It has given me great pleasure to hear that the great object of your going to Scarborough, the improvement of Frances's health, has been so far attained, and that you have derived benefit from the bracing air and sea water. I do not believe that you will have exhausted all the interest of the collections open to you, even at the end of another week. I had no idea that there were such riches in the way of collections at Scarborough as those you describe. I did not know of anything more than the Williamson's museum. I hope you will be able to add to your own collection some good characteristic specimens of fossil plants. I have got a new book from the London library which interests me, Kenrick's "Ancient Egypt under the Pharoahs." He is a Unitarian minister and schoolmaster at York, has the reputation of being an excellent scholar, and besides many other

good things he has written in reviews and magazine, is the
author of an admirable essay on Primeval History.   In this
new work, evidently the result of great labour, he has brought
together in lucid order the researches of all the great discoveries of Egyptian historical documents, and I doubt not
that his book will be ranked for a long time the best history
of Ancient Egypt in our language.   His spirit is quite liberal
and unfettered.

We had a pleasant visit to the Moore's and by the good
luck of bad weather, spent a most agreeable evening with the
Richard Napiers at Fir Grove.   We set out with the intention
of taking luncheon with them, but it rained so heavily that
they proposed our remaining, which we did, till next day,
most willingly.   They were both very well and *very* agreeable.

I am, my dear Charles,

Affectionately yours,

LEONARD HORNER.

To his Daughter.

Bonaly, *September* 13*th*, 1850.

MY DEAREST MARY,—You have probably heard of my arrival
here.   I have found Cockburn remarkably well, less marked
by advancing years than most people I meet with, whether as
to looks, vivacity, or freshness of mind.   I have had much
interesting talk with him about Jeffrey's papers, and have
seen some of his earliest productions, when he was only seventeen years of age, which shew a precocity of information,
philosophical reflection, taste, and power of writing, that is
truly wonderful ; even his handwriting was then formed, for
I do not think that anyone would be able to distinguish what
he wrote in 1790 from his letters in 1849.   There is an overwhelming mass of materials, it was his habit never to destroy
any paper, but at the same time they were thrown together
without order or arrangement of any kind, so that the task of

getting them into a chronological series has been the first task. It was rendered, however, comparatively easy by another habit he had, viz., that he dated every paper he wrote. There are recollections of his boyhood, of his feelings when, at eight years of age, he went to the High School, and his progress there, which are most curious and remarkable.

Among other places I visited yesterday was Mr. Bryson's shop, where I saw his son, who is a good mineralogist. He has been engaged in slitting into thin transparent plates, specimens of fossil woods ; among others psarolites, and is in communication with Robert Brown about them.

My dearest Mary,
Your affectionate father,
LEONARD HORNER.

To his Wife.

Carlisle, *September 16th,* 1850.

I breakfasted at Bonaly this morning, have been to Glasgow, have inspected three factories here, have dined, and have sent a note to Dr. Tait, proposing to sit an hour with him this evening.

My visit to Bonaly has been a very agreeable one and I think Cockburn was pleased that I had come to see him. He was very cheerful, and very playful, and we had much good conversation on many topics, much you may suppose about the Memoir of Jeffrey, about which I said something in a letter I sent to Mary on Saturday, and which I asked her to send to you. The world does not know what a remarkable person Jeffrey was, and his great merits and extraordinary powers were known to comparatively few. Had Jeffrey written one half with his name of that which was published anonymously in his reviews, he would have had an European reputation at this time. I do not doubt that we shall have two, or even more, very interesting volumes.

Mrs. Cockburn was as usual most hospitable and kind.
On Saturday I called on Lord Dunfermline at Colinton
House, and sat an hour with him.

To his Daughter.

Carlisle, *September 18th*, 1850.

MY DEAREST KATHARINE,—A letter which I wrote to your
mother will have informed you of my progress hitherward. I
dined with Dr. Tait yesterday, Mrs. Tait was confined on the
7th, so I did not see her. I hear but one account of the Dean,
he is universally popular. One man said that there had not
been such a Dean of Carlisle for many years, so active,
exemplary and benevolent ; taking a great interest not only
in Church schools, but of those of every other denomination. It
gave me great pleasure to hear this. The weather is beautiful
and very favourable for my moving about. God bless you,
darling, with love to Harry, and Nora, whom Mary tells me
was to be with you.

I am, your affectionate father,

LEONARD HORNER.

To his Wife.

Burnley, *September 26th*, 1850.

Your letter, received this morning, brought me the first
intimation of the sad illness of our very dear friend.* It must
have been little more than two hours after I parted with him
in the best health and spirits, that he had the attack. I
trust in God that Archie's account of his improved state has
been confirmed, and that he is getting quickly better.

29th September, Manchester. You will all be greatly grieved
to learn that I received this morning a much worse account of
our dear invaluable friend. His sufferings must have been

° Lord Cockburn.

dreadful. What an awful time for poor Mrs. Cockburn, and all his family. God grant that he may yet be spared to continue to bless them, and spreak joy around him as he has so often and so long done.

I can well understand your liking the vigorous writing of Theodore Parker, and his earnest search after truth, guided by a pure and lofty religious feeling.

To his Wife.

Manchester, *2nd October,* 1850.

You will read the enclosed letter from Archie Cockburn with great pleasure, for it justifies us in having strong expectations that all will yet go well.

I trust that you got safely to town. I am very glad you are there. I am sure you will sleep sounder for being so near our dear Kate. I am sorry you thought it necessary to go so much into detail about your cash transactions. You know very well that I know you to be the most careful, the most prudent, the most economical of wives.

Yesterday I was all day in Manchester, partly because a day of rest from visiting is better for me. As far as I have yet seen, there appears to be a very general contentment both among the masters and their work people with the new Act. The stopping at two on Saturday is much valued by the workpeople. One man said to me to-day, at Oldham, " Why, sir, it is like a little Sunday to me."

To Lady Lyell.

Manchester, *3rd October,* 1850.

I have heard nothing of Mr. Rogers since I left home, I hope he is going on well. I have nothing to say about myself as I lead a very uniform life, unproductive of events. One thing of an interesting kind I can tell from the factories,

There has been a terrible mortality among the silkworms, both in Italy and France. They were too much in advance of the mulberries and as there were no leaves, millions were starved to death, thence raw silk that a year ago was 10s. the pound, is now 19s., hence factories have ceased working, hence the workers are starving for want of mulberry leaves. I have begun Mackay's "Progress of the Intellect, as exemplified in the religious development of the Greeks and Hebrews," and am much interested. It is evidently the work of a man of powerful mind, of the most extensive acquirements and of a truly philosophical spirit. But it is certainly not light reading, although in general he is very clear. I am very curious to know who he is. I cannot believe that either Cambridge or Oxford can have produced him, and I doubt whether either place has, or for a long time has had, so accomplished a philosopher, or at least one who has had the courage to give utterance to his opinions as this author has had. It is another damaging book to priestcraft.

Friday morning. Oh! my dearest Mary, what a calamity has befallen the dear Nicholsons. I saw yesterday in a news-paper that George Henry Nicholson was one of the passengers in the Diligence. It will indeed be a very, very great blow, especially to his poor father, who I know looked up to him as a source of much happiness. God bless you.

To his Wife.

Manchester, *4th October,* 1850.

I sent off a letter to Mary after receiving yours. Poor Henry Nicholson and his bereaved parents and brothers and sisters have scarcely been out of my mind. . . . . Thank God, my eyes do not fail me, and I have books that interest me much, particularly that new work of Mackay's, parts of which you will read with great interest, others perhaps you will find less attractive. I have had a kind letter from Mallet, inquiring about Cockburn. He has seen Mrs. Marcet lately, and says that she is looking better than she has done for some years.

God bless you, darling.

Your affectionate husband,

LEONARD HORNER.

To his Wife.

Manchester, 13th *October*, 1850.

I have just returned from Mr. Tayler's chapel, where I have passed two hours very profitably and agreeably, with a simple and rational liturgy, very beautiful hymns, and an admirable sermon bv Mr. Tayler, on reliance on the wisdom and goodness of God. Before I went, I had the happiness to receive your dear letter. I had a good deal of talk with Mrs. Schwabe, who told me two cases of benevolence, in which she has been recently engaged, and successfully, shewing certainly great earnestness, activity, and sagacity on her part. They are both instances of the reclaiming of drunken husbands. She shewed me a beautiful letter of Cobden's on the death of Peel, which does him infinite honour. I have read Mr. Prescott's review of Mr. Ticknor's book; it is very interesting and makes me desirous of finding time to read the work.* I have read for a second time, the review of the " Second Visit " † and am even more pleased with it than I was at the first reading. It takes a long time to read for it is very difficult to resist the suggestions it constantly gives rise to, and I find my thoughts constantly wandering into speculations of what advantages would accrue to our own country if the principles of religious freedom, extended education, and an extension of the

------

° The "History of Spanish Literature."

† Second visit to America by Sir Charles Lyell.

democratic principle (when by education it could be safely done) so ably expounded, were adopted in our land. The article is calculated to do much good in this country, and in the latter part of it especially, in the United States I trust that it will have the good effect of impressing on the American people a conviction that a very considerable portion of the people in this country take a deep interest in their welfare, and would gladly follow the good examples they have set us in very many things. It is no small part of the good that Charles may be gratified in feeling he has brought about, by the time and thoughts he spent in that work, that it has given birth to so admirable a production as Spedding's review.

———

To his Daughter.

Manchester, *October* 17*th*, 1850.

MY DEAREST LEONORA,—What charming letters those are of Mrs. Ticknor's. She speaks of our dear Mary as she would do of a dear loved daughter. It is very delightful that Susan's portrait* is so much liked, and gives so much pleasure. How happy Mr. Prescott must be, and how they will hang upon his narratives of all he has seen. I find his volume of Essays very agreeable reading. I had not read that on Walter Scott, when Mary's recommendation of it came. I read it the same evening, and was much pleased with it. He rightly understands and appreciates Scott's character. His style is very agreeable. I have since read the Essays on Cervantes, and Irving's "Conquest of Granada," both of which are very good—very instructive.

You say that you had been to the British Museum. There has been some good articles in the *Athenæum* and *Examiner*, upon the crowded state of it. The time cannot be far distant when a great public library will be built, and there ought also

* Medallion portrait of Lady Lyell.

to be a Glyptotek so that all the Museum should be given up
to Natural History.

I admire Dr. Holland's spirit greatly. There are not many
men in any profession who have made such good use of their
vacations as he has done.

From T. G. Nicholson, Esq.

Waverley Abbey, *24th October*, 1850.

DEAR MR. HORNER,—How delightful it is to think of the
length of our friendship, undisturbed as it has been by any
of those circumstances which are so constantly interfering
with the connections of this life! and how pleasant it is to me
to think that this friendship, will be a legacy to our children,
which they seem well-disposed to value. Thank you for
reminding me of it, and many thanks for adding, by your
kind sympathy, one instance more of the lively interest you
take in everything that concerns me. You and Mrs. Horner
know full well the way to a parent's heart. Your uniform
kindness to my children, and to him I have lost in particular,
has always been most gratifying to me, and now that I know
how deeply you appreciate poor H.'s character, at the same
time that it aggravates my loss, still deprives death of half its
sting. I can say no more upon this still painful subject. I
was sorry to hear that you had returned home uneasy about
your legs; I hope the inconvenience will be soon removed,
and that nothing will interfere with your joy at your daughter's
safety, and the birth of a grandchild.

Believe me, my dear Horner,

Very sincerely yours,

G. T. NICHOLSON.

To his Daughter.

Rivermede, *10th November*, 1850.

MY DEAREST FRANCES,—I am rejoiced that you are in such
good heart as to the ultimate success of your boys' school. I

have been in some correspondence with Mr. Zincke, and you will read with pleasure what he has said in his forthcoming pamphlet, of the King's Somborne School, and of the Mildenhall School. The death of Henry Hallam is indeed a most calamitous event. I doubt whether his poor father will have strength to bear this fresh blow. I have not heard that any account of them has been received since a letter written from Florence.

It is a great blessing that our dear Katharine has made so good a recovery, and that the dear boy is thriving. 1 trust that our scheme of visiting you *in a body* at Christmas will be realized, but I cannot understand how you can possibly receive us, unless you make the long gallery into a dormitory. If you are not afraid of the visitation, I am sure every one will be delighted with the reception given.

I do not doubt that you and Charles have read with great pleasure Lord J. Russell's admirable letter to the Bishop of Durham. It is worthy of a Russell. The spirit that is now roused will I hope give a blow to Puseyism from which it will not recover. I have observed that in not one of the denunciations of the clergy against these inroads of Popery, has any mention been made of the worst of all their proceedings, viz., the interference with the Queen's Colleges in Ireland by the Synod of Thurles. Our prelates and parsons are thinking only of the assumption of their own tithes by the Papists. They are evidently trying to turn the Pope's movement to their own account.

My love to Charles.

Your affectionate father,

LEONARD HORNER.

To his sister, Mrs. Byrne.

Mildenhall, *27th December,* 1850.

MY DEAREST FANNY,—You will believe the very great happiness that my dear Anne and I have, to be surrounded with our six daughters, our three sons-in-law, and our darling grandson. It is a blessing for which we cannot be sufficiently grateful, but we must not forget the uncertainties of life, and be not so elated as to forget that sad changes may be not far off, and keep that sobriety of joy which, without damping the present, keeps the mind in a better tone for the possible, nay probable future.

We had " a merry Christmas," but yesterday was the great day. Besides our own family, Edward Bunbury and his nephew Harry are here; about one o'clock Pellew* arrived with his daughter Fanny and his two youngest sons. About five o'clock we were all assembled in the drawing-room. Baby was dressed in *the* white satin frock, the history of which was fully told.† Mr. Pellew read a short and simple service from the Prayer Book, took the dear little fellow in his arms, and baptized him " Leonard." The young gentleman behaved perfectly well until the water was sprinkled upon him, and then he called out lustily, which his nurse said was a good omen. We then dined, a merry party, and immediately after we adjourned to the book-room, where a cake surrounded with fruits, wine, etc., enclosed in a circle of candles, had been arranged by the girls. We had then a variety of toasts, succeeded by games, and much merriment, a round game closing the joyous evening. All are well to-day, and little Leonard paid me a visit in my room, and smiled and crowed merrily.

We paid a most agreeable visit at Waverley,‡ and I believe we did them some good, by their being able to talk to us a great deal of their poor lost Henry. Hannah was, as she always is, very delightful, my heart warms to her as to a

---

° The Hon. and Rev. Edward Pellew.

† A relic in the family, it having been the christening frock of Francis and Leonard Horner, as well as their mother, who was baptized by Dr. Robertson.

‡ Waverley Abbey, the seat of G. T. Nicholson, Esq.

sister. We passed four very pleasant days at Embley,* and during the time I went on two successive days to meet Mr. Dawes, the Dean of Hereford, in his far-famed school at King's Somborne, the living of which he has held for several years, and quits with great regret. I have seen many schools, but no one with which I have been on the whole so well pleased. He has, as you perhaps know, carried out the great principle of bringing together the children of the labourers and of their employers, with complete success. I found him quite acquainted with the letter of our dear Frank describing his visit to Mr. Poole's school at Enmore in 1812. I was quite delighted with the attainments, the intelligence, and happy, cheerful looks of the children, boys and girls, the perfect confidence they showed in their teachers, and their cleanliness in person and in clothes. Their knowledge, both for extent and accuracy, was quite surprising, and their singing beautiful. It would have done your heart good to hear " The Boatie Rows," set to very pretty words, and " Ye Mariners of England." I have struck up an intimacy and active correspondence with Mr. Dawes, I had the gratification to hear from him that my translation of Cousin, and my Preface, were among the first things that called his attention to education. He is a most admirable, enlightened person, and if there were one thousand such as he, the Church of England and education would be vastly improved, beyond what either is at present. I do not know whether you ever saw the pamphlets he has published, in one of them there is an excellent account of his visit to the National Schools in Ireland. There is an awakening in the Church itself, a considerable spirit of advancement in education and freedom of enquiry, which should meet with every encouragement, for the clergy are so numerous, that they have enormous power to retard or advance education, and every step in the right direction is a gain, and brings us

* Embley Park, Mr. Nightingale's country house.

nearer to that dominion of truth which I believe the world is destined to see, when priestcraft of every denomination, and all priestly domination, shall be swept away ; and it may be one of the joys of heaven to witness this purging of the nether world.

We live at a moment of extraordinary excitement in this country on that same subject of priestcraft, the attempt of the Catholic priests to gain additional power, has roused a spirit of intolerance among the vulgar, which one had hoped had passed away for ever. But the attempt has been so unblushing and audacious, has been so wholly uncalled for by any restriction upon the most perfect liberty to worship God in the way that members of that sect of Christians liked best, that on the ground of protecting intellectual freedom and civil liberty from great danger, it must be, and I trust will be, manfully resisted. What are the legislative measures that can be adopted consistently with entire religious toleration, it is not easy to guess. It is, I think, clear from all one reads of the large majority of the speeches and resolutions throughout the country, that no class of the community desires to see any retrograde movement in regard to religious freedom. I do not believe that the movement will strengthen the English Church, as respects the power of the priesthood; on the contrary, I believe it will lead to a weeding out of the Athanasian Creed, and all other remnants of the Roman Catholic superstition, the tendency of which, and designed end of which, is to bring the understandings of the laity under subjugation to the priests. Puseyism at all events has got a blow from which it will not soon, if ever, recover. How much we miss Sydney Smith at this time! what inimitable combinations of sense and wit we should have had from his pen.

I am, my dearest Fanny,

Your affectionate brother,

LEONARD HORNER.

CHAPTER VIII.

1851.

To his Daughter.

Rivermede, *7th February,* 1851.

MY DEAREST FRANCES,—It gave me much pleasure to learn
from your own letter, and from that of Charles, how very
satisfactory the visit of the Dean of Hereford has proved, in
several ways. He is indeed a most agreeable person, so cheerful, so sensible, so rational, so benevolent, so kind, and then
so enlightened. I am very glad indeed that he has been with
you, for it gives you courage and confidence in your plans for
your schools. The visit cannot fail to tell in Mildenhall. It
will give great support too, to Mr. Phillips and Miss Scott.
I think that the gradual advance to twenty in the boys' school
is very encouraging, and when the fine weather comes, and
their joyous voices are heard in the playground, the popularity
of the school will doubtless be increased.

As soon as I got Charles' letter, I filled up a recommendation for the Royal Society, and sent it to Brown, that
the name of the *Princeps Botanicorum* might stand at the head
of his proposers. I got it back this evening with a note
saying that he had signed the paper with great pleasure. I
have sent it to Sir William Hooker, that his name may be
next, and I shall apply to the other eminent botanists in this
order :—Lindley, Forbes, Royle, Wallich, Daubeny, Stokes,
Lankester, Horsfield. With these names he will have the
support he so well deserves.

My kindest regards to Charles,

I am, my dearest Frances',

Affectionate father,

LEONARD HORNER.

To his Daughter.

Manchester, 16*th March*, 1851

MY DEAREST KATHARINE,—I am very much interested in all
the accounts that come to us about your darling boy. He
seems to be thriving famously, and to be growing more intelligent daily. How much I should like to see him!

I hear from most quarters a very favourable account of the
prospects of the educational plan here. I had a conversation
yesterday with Mr. Barnes, a leading man among the Wesleyan
Methodists, a mill-owner, who is a member of the Committee,
and who subscribed £500 towards the fund for carrying the
measure through. I hear, too, that there is good ground for
expecting a cordial co-operation from the Roman Catholics;
their clergy here are exceedingly well-disposed.

I have read the whole of Sedgwick's discourse, and was
much interested by it. There is a great deal of valuable
information; but this book written in a spirit of honesty from
beginning to end, I am persuaded, is a proof how blighting to
the best understanding it is for a man to be brought up to a
profession that he is sworn to maintain. If Sedgwick had not
been a churchman, we should have had very different opinions
on many subjects from him, than he there maintains. But
he is not consistent, he tells us that Paley's " Moral Philosophy " is the text book for that department of knowledge, and
he condemns its principles in the most unqualified terms; yet
not a word in condemnation of its being placed in the hands
of their youth, and he does not recommend, nay, not even
suggest, another. He does not allude to Stewart—so much
for University prejudice, Love to Harry.

Ever, my dearest Katharine,

Your affectionate father,

LEONARD HORNER.

To his Daughter.

Manchester, *April* 5*th*, 1851.

MY DEAREST MARY,—Thank you for letting me see Mr. McIlvaine's interesting letter. It must be a great pleasure to you that you have such excellent and steady correspondents to keep up a familiar acquaintance with the passing events around your American friends.

I had a few lines from Colonel Sabine on Tuesday, telling me that Hopkins had reported favourably of my projected experiments in Egypt. When I was at Embley last December, Florence Nightingale, to whom I spoke about my researches, said that there is a Mr. Harris in Alexandria very likely to assist me ; she gave me a letter for him which I forwarded, explaining to him my views. I did not expect an early answer, for Florence said he was in Upper Egypt. I got yesterday two letters from him, one dated the 11th, the other the 19th of March. He enters most warmly into my plan, and has engaged a very intelligent engineer in the Pacha's service, who was educated in England ; he superintends the boring. He sent me a copy of a letter this engineer had written to him on the subject, which is highly satisfactory. He, the engineer, had spoken to Abbas Pacha, the Viceroy, on the subject, as his Highness has a garden at Heliopolis in which the obelisk stands, the spot, or at least one of the spots, most adapted for the experiment. He found that the Viceroy had already been spoken to by Mr. Murray, and he has given full authority to the engineer to sink the shaft at the place most desirable, and according to such directions as he may receive from me. The engineer says that June is the best time to sink the shaft, so that I am still in good time for this year. My absence from London just now is unlucky, as I should have been glad to have consulted Bunsen personally on many points before giving the final directions, but I have written to him, and have sent him Mr. Harris's letters. . . . . . .

I am, my dearest Mary,

Your affectionate father,

LEONARD HORNER.

From Lord Cockburn.

Glasgow, 8*th April,* 1851.

MY DEAR HORNER,—I have just got your letter *here*, where I am on my *circuitous* way to Inverary. I shall get to Bonaly on Saturday first, the 12th, in the evening; and shall rejoice to see you at Bonaly on Sunday to breakfast, dinner, tea and bed. Pillans accompanies you of course. I can put nobody but you up at night. But you I expect to stay at least over the Monday.

And *take special notice,* that your going to live with Pillans is a gross breach of ancient and exclusive allegiance to me; and a piece of dirty, ungrateful, School of Arts puppyism, and I am strongly inclined to give you up. You will find me quite well. All our hearts are in mourning for Moncrieff. Certainly last September I never thought that he was not to mourn for me.

Love to Mrs. Horner, *who would not have deserted me* for an old, fozzy, semi-doited, male, widowed, professor.

Perfidy, thy name is Leonard !

But still I remain,

Thine ever,

H. COCKBURN.

To his Sister, Mrs. Byrne.

London, 30*th April,* 1851.

I have been so very much engaged, my dearest Fanny,

for the last two months that I have been unable to
write to you, or to Nancy. My visit to Edinburgh
was productive of pleasure in many ways. I was invited
to come by some of the directors of the School of Arts,
as there was a negotiation going on about the house it has
occupied for fifteen years, and there were some difficulties
which they thought I might help them to overcome. This
appeal to my affection for my old friends was not to be
resisted, and they were pleased to say that my visit has been
of use, for the negotiation was brought to a satisfactory
conclusion. The house is the centre one in Adam's Square.
The intention is to have a new architectural front, and to
have a copy in Craigleith stone of Chantrey's sitting statue of
Watt, in the Square, and these two combined to be the
monument in honour of Watt. Nothing certainly could be
more gratifying to me than to find the School so flourishing,
and so generally held to be one of the established educational
institutions in Edinburgh.

I found Cockburn very well, thinner somewhat, but very
cheerful. He is very much occupied with his Memoir of
Jeffrey, and has advanced as far as the year 1829. He
intends that there shall be a volume of Memoir, and *for the
present*, a selection of letters sufficient for another volume.
As far as is wanted to illustrate the character of Jeffrey, the
materials are abundant, but there were few incidents in his
life to supply a memoir that will be very interesting to
strangers. Those who knew him could hardly over-estimate
his excellence, his genius, his manly, noble independence, his
fancy, his acquirements, but he was wonderfully little known
in England, scarcely at all to the great mass even of educated
people.

I lived with Pillans, in his pretty and pleasant house in
Inverleith Row, close by the beautiful Botanic Garden. He is
remarkably well, cheerful and vigorous. He is up every

morning between five and six, and walks out before breakfast. He is seventy-three, and is as active as a man of fifty.

I met Tom Jackson* and his wife, both very agreeable. He has acquired a very high reputation as Professor of Divinity at St. Andrews, and has just been appointed to the more lucrative office of Professor of Church History in the University of Glasgow. There are few men of the Church of Scotland so learned or so liberal. Rutherfurd I sat half-an-hour with, and found him looking better and less feeble than I expected he would do, after so very dangerous and protracted an illness. He has been fortunate in having had an opportunity of being placed on the Bench, for he could not have gone on with the laborious duties of Lord Advocate.

1st May. A glorious day, and the opening of the Exhibition has taken place with perfect complete success, and Lyell says it was the most magnificent pageant he ever witnessed. At eight o'clock Lyell and Mary, Joanna, Katharine and Harriet Moore went off; Charles, in his scarlet uniform of a Deputy Lieutenant, having to take his place as a Royal Commissioner in the Queen's procession. They say nothing could exceed the splendour of the sight, and not a thing went wrong. The Queen and Prince performed their parts admirably, as they indeed always do. Nothing could be more perfectly orderly, or in more perfect good humour than the dense mass of people that covered the park, and the enthusiastic loyalty to the Queen and Prince, so well merited, must have made a deep impression on the multitude of foreigners. Another thing must have struck them, that all the admirable order and arrangement of the day was kept up by the police in plain clothes, that is their plain uniform, without arms of any kind, and very few soldiers were to be seen. Russia and Austria refused passports to their subjects to come to the Exhibition ! !

* Nephew to Mr. Pillans.

Give our kindest regards to your husband and the Powers,
and

Believe me, my dearest Fanny,

Affectionately yours,

LEONARD HORNER.

To his Daughter.

Rivermede, *May* 18*th*, 1851.

MY DEAREST KATHARINE,—It was a great joy to receive your
dear letter from Drumkilbo,* I am most happy to learn that
you had a tolerably good voyage, and that our darling boy
was not the worse for it.

As Mary wished to know which of your sisters was to go
with them to Germany, I had to decide upon our going.
Joanna is "to follow the Newer Eocene" with Charles and
Mary, and Susan and Nora go with your Mamma and me.
We shall set out within a few days of your departure,† and it
will be some diversion to our minds, for we shall need it much.
We talk of going from London to Antwerp by steamer, then
to Bonn, Wiesbaden and Frankfort, and proceed to Dresden,
then to Nürnberg and Munich and thence to Switzerland by
Innspruck.

My love to Harry, and a kiss to the dear boy.

God bless you, darling,

Your affectionate father,

LEONARD HORNER.

To his Daughter.

London, *July* 12*th*, 1851.

The first packet for my dearest Kate must not go without a
contribution to it from me. It was an unspeakable pleasure
to receive your dear letter, written an hour before you sailed,
with so cheerful an account of your first night on board;
and the kindness of Lothian Nicholson in going on board

° In Perthshire, staying with her sisters-in-law

† For India.

and staying with you for some hours, I never can forget.   I have watched the weather, and it has been fine since you left us. Your account of your darling boy is very satisfactory.   I think of the last look I had of the dear fellow at the window of the railway carriage, with a sweet smile showing his four pearls.

We are all ready to set out to-morrow ; this is a fine day, and as the wind is in the west just now, we may hope for a good passage to Antwerp.   Mamma, Susan and Nora were to go to the Crystal Palace to meet Count von Randwyke, who was with us this morning at breakfast time, and they were to go to the concert for the Hungarians afterwards, and to hear Fanny Kemble read.   I could not accompany them to either, for I had to go to the City, and to the Factory Office.

The grant of £150,000 for education was made last night, and Lord John Russell gave some hopes that there would be greater activity on the part of the Government on that subject.   There is a story to-day in the newspaper that Rutherford is to be consulted by the House of Lords in appeals.   Whether there be any truth in the story or not I cannot say ; but as he is said to be a very good lawyer, he may do much good, not in matters of appeal, but in very many questions connected with the law of Scotland.

I daresay you have been told much about the Bunsen's party last night, and how heartily Count and Countess Beust met us.   I do not doubt too that you have been told of our having had a letter from Joanna, dated Calais, yesterday, where they had safely arrived.

God bless you all three, I trust this will find you much improved in health by the two months of sea air and sea-water.   Kiss my sweet pet for me, and remember me to Ayah.

My dearest Katharine,

Your affectionate father,

LEONARD HORNER.

To his Daughter.

Munich, *July 28th*, 1851.

My DEAREST FRANCES,—Nürnberg is more peculiar in its
character of domestic architecture than anything I ever saw,
and the profusion of projected ornamental parts has a fine
effect in the general appearance of the streets. I never saw a
place where the houses appear to be kept in such good repair,
and where cleanliness in the streets is more observed. We
left Nürnberg on Saturday about one o'clock, by the railway
which brought us to Augsburg by seven, where we put up at
an excellent inn, *Drei Mohren*, which has been an inn for three
hundred years. the house having belonged to the great mer-
chants, the Fuggers, who entertained Charles V. in it. The
room where the banquet took place, and the stove where he
(Fugger) made a fire of cinnamon, to burn a bond for a large
sum Charles owed him, were shewn to us. We went with
much interest to look at the place where the Confession of
Faith was read to Charles, but the room no longer exists.
The Maximilian's Strasse is one of the finest I have ever seen.
We looked into the cathedral, where mass was going on, and
there was a dense mass of people filling the church. We left
Augsburg at half past one, and were here at four. In the
evening we drove to the Theresiens Wiese, to see the colossal
Statue of Bavaria in bronze by Schwanthaler. It is above
fifty feet high, but the most beautiful softness of expression of
the countenance is given, and the drapery is beautifully
executed. It is a noble work of art. We have been this
morning to the Pinacothek, a most valuable collection. This
is certainly the most beautiful town I have seen, and excellent
taste is conspicuous everywhere.

All join in affectionate love

With your fond father,

LEONARD HORNER.

To his Daughter.

Zürich, *August* 10*th*, 1851.

My Dearest Katharine,—Often and often have I thought of " the Maidstone " and its precious contents, and speculated about each of you. Most fervently do I pray that when this letter is opened, you may be safe and well on terra firma, and improved in health by your long voyage.

We have enjoyed ourselves much, and though our plan has been partially changed by bad weather, the alterations have been accompanied by considerable pleasure, above all by our having seen the Fall of the Rhine at Schaffhausen, which in grandeur of *effect* and picturesque beauty much exceeded my expectations.

Our stay at Bonn was very pleasant, and I should have been glad if it had been possible to prolong it, for we had more persons and things to see there, than we could do, and to have dwelt with some quiet leisure among them would have been very pleasant.

I have received a letter from Mr. Harris of Alexandria, with copies of letters he had received from the engineer Hekekyan Bey, written at Heliopolis, giving an account of the progress of the excavations there, which began on the 10th of June, and had advanced considerably, when he last wrote, on the 28th, and with good promise of interesting results for the object I have before me, in getting these researches made. He has dug up some architectural remains which promise to be of considerable interest.

Besides the pleasure derived from the journey, I have been *sehr fleissig* in trying to improve my German, and hope by the time I get home to find a great improvement both in the power of speaking and reading.

I have been reading for my travelling study Wilhelm Meister's *Lehrjahre*, which I had not done before. I am considerably disappointed in it, and Goethe is lowered in my estimation.

To dwell so much upon a set of wretched strolling players, who shew so little genius in their art, is tiresome, and his hero is a feeble personage. There are some amusing incidents, and some good criticisms and reflections. There is a sensuality in several. places, in no way necessary to the story, which indicates a coarse mind in the author. I do not think Schiller could have written in such a style ; it would have been quite contrary to his loftier and purer spirit.

God bless you, my dearest Katharine, my affectionate love to Harry, and a thousand kisses to your darling boy.

Your affectionate father,
LEONARD HORNER.

To Sir Charles Lyell.

Vevay, *August 21st,* 1851.

MY DEAR CHARLES,—Here we are on the beautiful Lake of Geneva, and I believe at one of its most beautiful parts. We have made our first acquaintance with it under most favourable circumstances, for we have had a very fine day. We got the first sight of the Savoy Alps about ten miles from Vevay, when we saw the Dent du Midi's snowy top above a mass of fleecy clouds. Soon after we began to descend, and had the beautiful valley that rises behind Vevay before us all the way, until our arrival. We got here just as the second table d'hôte at five was commencing. It was very good, for we are at a capital hotel, *les trois Couronnes*, and did not last very long. As soon as it was over, we rowed for an hour upon the lake, when we had a magnificent sunset, the Dent du Midi brilliantly lighted up, of a bright red, and a rosy tint upon the summits of all the neighbouring mountains ; the lake smooth as glass, and like a sheet of fire for a vast distance. We saw Chillon and the places celebrated by Rousseau and Byron immediately before us. Had we arrived sooner, we should have gone to Chillon, but if we had we should have lost the charming row on the lake during sunset. As we were descending to the town we had a fine view of the great plain, the embouchure of the Rhone. We did not find much to interest us at Berne, the Oberland Alps were not to be seen from the terrace, being enveloped in clouds. We started for Fribourg after the early table d'hôte, and got there with light enough to see the beautiful and wonderful suspension bridges, for there are two. The situation of the town is most picturesque. The great organ is played every evening at seven, and we went of course to hear it, which we did for an hour, and were astonished by its power, and charmed with its sweetness. It was admirably well played. Travelling as we are now doing, all geological observation is out of the question. But I have seen enough to make me lament that I am not thirty years younger, with time and a pair of good legs. I was much struck with the vast extent of the gravel plain that stretches all along the foot of the Tyrolese Alps. I crossed it for forty miles from Munich, and traced it for more than seventy miles towards the lake of Constance. It looks all along like fresh water gravel. I have been still more struck with the enormous deposits of the Nagelflue, and its upheavings. I think I saw it first in the Albis crossing from Zurich to Lucerne, and I have seen it again to-day with the same characters within a few miles of the lake of Geneva. What a tale it tells! I did not fail to look with wonder on the contortions of the strata on both sides of the lake of the four Cantons, between Brunnen and Fluelen. I cannot bring myself to believe that the vast, precipitous, ragged and gagged masses with which I was surrounded in that region, can have been produced by any other than a sudden *fracture*. In crossing the *low* country from Thun to this place, and observing the hills we were constantly ascending and descending and the numerous crags and escarpments I saw on all sides,

I thought of Agassiz's theory of the movements of Glaciers across the so-called plain of Switzerland, between the Alps and the Jura, and wondered how such a notion could enter the mind of any one. I believe Forbes too entertains the same view. It is to me utterly inconceivable.

I am very happy that you are finding so much to interest you in Belgium, and that you will be able to set the specimen collectors there, on a right tract. If you do no more than that, you will have done good service to science. But I expect that next winter the Geological Society will benefit much by your present researches.

We intend to be at Lausanne early to-morrow, and go on by the afternoon steamer to Geneva.

I am, my dear Charles,
Affectionately yours,
LEONARD HORNER.

To Captain Henry Lyell.

Rue de Moncean, Paris, *September 5th,* 1851.

MY DEAR HARRY,—We often and often think of the dear triad on the ocean, and although anxious thoughts will force themselves, we do our best to believe that all is well, and that the dear boy is thriving as perfectly as can be desired. We hope that you have by this time doubled the Cape, as you have now been more than eight weeks at sea.

When at Chamouni we all started for the Montanvert and Mer de Glace, your mother and I in chairs carried by two men, with one relay for her, and two relays for me. Susan and Leonora preferred walking a part of the way, so they had a mule between them; Mr. Brandis walked, but his son John rode a mule, as he is delicate. We were a party of seventeen. It took two hours and a half to ascend, but it was not at all difficult, and we descended upon the Mer de Glace. We were all, you may suppose, lost in admiration of the beautiful and

wonderful scene before us. We reached Chamouni without
accident of any kind, much satisfied with the feat we had
accomplished. On the evening before, Susan, Nora and I
went to the Glacier de Bossons. We afterwards saw the
source of the Arveyron, pouring forth from the Glacier du
Bois, under a great vault of clear blue ice. We left Chamouni
that afternoon, the 27th, and slept again at St. Martin. We
got to Geneva next day to dinner. On the 29th we dined
again with John Prévost at Bouchet, and met his brother,
Frank Marcet, and William Romilly. The latter we had not
seen for thirty years and more. We left Geneva very much
pleased with our visit, and admire the place much. We went
in a diligence on the 30th from Geneva to Lyons, starting at
six o'clock in the morning and getting to Lyons at half-past
seven, ninety miles, capital driving, through a fine country,
and the road crossing the Jura in many places very grand.

We started at five in the morning of the 31st in a steamer
for Chalons, up the Soane, got there soon after one, and then
were carried in two hours more by railway to Dijon. Dijon is
a very pretty town. On Sunday morning at eight we started
by railway for Paris and reached it by half-past four, two
hundred miles, a capital railroad, first rate engineering, and
the most comfortable first class carriage, a *coupé*, I ever
travelled in. We found my sisters and their husbands perfectly
well. They are all kindness and hospitality. We are
amusing ourselves with sight seeing, and yesterday at the
Hippodrome saw the most wonderful feats in horsemanship,
tumbling, posture-making and slack rope dancing, I ever
witnessed. The most extraordinary feat was that of a man,
who, *standing* upon a broad wheel about two feet in diameter,
rolled it up a plank and down again, the plank about two feet
wide, being elevated at an angle of thirty degrees and about
seventy feet long. He did the same with a great ball, two
feet in diameter, standing upon it. It is certainly a most

beautiful *tour de force*, preserving the centre of gravity so perfectly, for the least departure from it would have hurled him headlong.

We intend to remain here until next Tuesday, when we shall sleep at Boulogne, and cross next day to Folkestone. On Friday we hope to be at Rivermede. After we have been a week at home, Mamma, Susan and I go to Mildenhall to see Frances and her husband before they go abroad for the winter; I shall stay a couple of days and then proceed to Lancashire, and when I have found quarters at Manchester, Mamma and Susan will join me.

Give my affectionate love to dearest Katharine, and kiss the darling boy for me.

I am, my dear Harry,

Yours affectionately,

LEONARD HORNER.

To his Daughter.

Manchester, *Ocotber* 13*th*, 1851.

MY DEAREST FRANCES,—I was most happy to hear that Bentley had agreed to publish your Dante. I wrote some days ago to Sir Henry to thank him for the gift of his late work, and for the pleasure the perusal had afforded me. It is a most admirably drawn up narrative, and his style is so pleasantly natural, and flowing, and cheerful, and spirited, that it was with difficulty I could lay the book down. The descriptions of the battle of Maida, and the taking of the Island of Capri by Murat's troops, are quite admirable. But what a picture it gives of the imbecility and ignorance of our governments in their plans, and the men they chose to carry them into effect, except Sir John Moore.

We have been, as you may suppose, sadly grieved by the death of our dear and excellent Mrs. Mallet. What poor

Mallet is to do without her tender hourly care, it is difficult
for me to imagine.

Give my affectionate regards to Charles, and with Mamma
and Susan's best love.

I remain, my dearest Fan,
Your affectionate father,
LEONARD HORNER.

To his Daughter.

Manchester, 16*th October*, 1851.

MY DEAREST KATHARINE,—We have seen in the newspaper
that "the Maidstone" had been twice spoken with, I trust
therefore that you have been making steady progress and that
you are by this time far advanced in the Indian Ocean.

Yon will greatly lament the loss we have all had by the
death of our dear friend Mrs. Mallet. . . . . Our newspapers are full of accounts of public meetings held for the
purpose of expressing sympathy with Kossuth on his arrival
in England, where he is daily expected. He is to be received
at Southampton by the Mayor and Corporation in their official
dresses, and a great banquet is to be given in his honour there.
The Pulszky's. and Colonel Gall are at Southampton waiting
for him. The disgraceful conduct of the President of the
French Republic and his ministers, not to allow Kossuth to
travel to England across France, has excited a general indignation in this country, and must be severely condemned by a
great proportion of the people of France, and the contrast of
England and France will be very galling to the liberal minded
among the French. The American Government has acted
nobly. There has been a great meeting at Boston to celebrate
the opening of the railroad which unites Canada, and the
United States, it being now open from Boston to Quebec.
Lord Elgin, the Governor-General of Canada came, as did the

President of the United States with several of his Cabinet. The speeches of Lord Elgin, Mr. Everett, and Mr. Winthrop are quite admirable. Mr. Winthrop is about to be elected Governor of Massachusetts.

I am happy to say that the Manchester scheme of Education for the boroughs of Manchester and Salford is going on most favourably, and if carried through, will do more honour to Manchester in my opinion than anything her citizens have done. They will have solved the two great difficulties that have hitherto impeded the progress of education—want of money and religious differences. The money is to be obtained by a school-rate, and *all* schools, of every religious denomina- tion, are to be assisted out of this rate, good secular teaching and inspection being the chief condition.

Your Mamma had a letter from Mrs. Cockburn to-day, giving a good account of them all. Lord Cockburn in excellent health and spirits. In a letter I had from him ten days ago, he says he has nearly completed his Memoir of Jeffrey.

Dear Frances is much pleased that Bentley has agreed to publish her translation of Balbo's " Life of Dante," entirely at his own risk.

God bless you, my dearest Katharine, and with love to Harry and many kisses to my dear grandson.

I am, your affectionate father,

LEONARD HORNER.

To his Daughter.

The Grove, Highgate, *November* 18*th*, 1851.

MY DEAREST KATHARINE,—Your dear affectionate letter to me from the Cape, dispatched the 21st of September, reached us a week ago, to our great joy, as we did hope that you would write again from the Cape. I fervently trust that Captain Nash's expectations were realized, and that you all landed safely at Calcutta in six weeks from sailing from the Cape,

which would be about the 3rd of this month, I fear that you
encountered a very rough sea for many days.

You are a perfect letter writer ; first you have the grand
element of a warm affectionate heart, then you have a vivid
picture before you of the anxious friends you are addressing,
and you go into all those details about your husband, your
darling boy, and yourself, and the people who constitute your
fellow passengers, which are of all things the subjects most interesting to us. All the minute circumstances respecting the
dear little fellow, his health, his development and growing
intelligence, we have read with peculiar pleasure, and we hope
most earnestly that you may be able to continue your satisfactory accounts. Your stay at the Cape appears to have
afforded you much pleasure, and I doubt not did you much
good in point of health. You were most active in your
botanical researches, and you will be glad to hear that we
heard a few hours ago from Mary that the plants you sent
home had come to Harley Street, from the Admiralty, forwarded by Sir F. Beaufort.

Well, here we are fairly established on the brow of Highgate
Hill, in the house Harry saw, and was so well pleased with.
That next to it has some celebrity as having been for many
years the residence of Coleridge, and he died in it. *Boreas*
has given us a hearty reception after his fashion, as it is not
in his line to deal in warmth. The day after we arrived
(Friday the 14th was the first night your Mamma and I slept
here) there was a hard frost, which has continued ever since,
and accompanied by much wind.

I have got my library nearly arranged and it is very comfortable. I think we have every reason to be satisfied with
our house. To-morrow I go to the Geological Society.

The electric telegraph under the sea from Dover to Calais
has just been completed, and is now daily in active operation.
The communication with Paris is now for all important intelli-

gence reduced to half-an-hour. In time I suppose the
Governor-General when at Lahore will hold conversations
with his Council at Calcutta.

Charles and Mary dined here on Friday, his birthday, both
very well. Charles as usual very busy. I have had several
letters from Egypt describing the operations going on there,
suggested by me, which are very interesting, and the first
leisure I can command I shall set to work to make use of
these new materials. Unfortunately Hekekyan Bey, the
engineer who has been directing the operations, had been
suffering from severe ophthalmia.

My best love to Harry, my blessing upon the dear boy, and
give him many kisses from grandpapa. God bless and preserve you, my dearest Katharine, is the earnest prayer of

Your affectionate father,

LEONARD HORNER.

To his Daughter.

The Grove, Highgate, *December* 16th, 1851.

MY DEAREST KATHARINE,—First let me congratulate Harry
on his brevet rank. I should be glad to hear that it gave him
any substantial advantage, or that it hastened, even by a day,
his return to England. The political news of the last fortnight
has been particularly exciting since the *coup d'état* of Louis
Napoleon on the 2nd, which amounts to another revolution.
His conduct has been the most daring and wicked that ever
disgraced any man ; perjury, lying, oppression, cruelty, everything disgraceful. He is supported by the army, and there
now reigns in France as complete a military despotism as in
Russia or Austria. What it will lead to, and what it will end
in, no one can predict ; but such is the feeling in France of
security under military rule, that their funds have advanced
within the fortnight nine per cent. You may conceive under
what a state of terrorism they live in Paris by this, that your

aunts, who are writing twice a week, have not ventured to allude to politics. Not only are all letters opened, but spies are placed everywhere. There seems every probability that Louis Napoleon will be re-elected President by a vast majority, and for ten years.

Joseph Hooker is now settled at Kew, working at his collections.

I am going on with my Egyptian researches. The excavations in the ground near Heliopolis have been carried on with great spirit by Hekekyan Bey, at the Pacha's expense, thanks to Mr. Murray, and I have had ten letters describing the operations, full of interest, and giving some curious results. Next year as soon as the waters have sufficiently subsided, they propose to make similar excavations on the site of the ancient Memphis, the city founded by Menes about 3,800 years before the Christian era.

December 18*th.* To-morrow the Edward Forbes's and Erasmus Darwin dine with us. Charles Darwin was at the Geological Society's Club yesterday, where he had not been for ten years—remarkably well, and grown quite stout.

God bless you. My love to Harry and kisses to the darling boy.

Your affectionate father,
LEONARD HORNER.

CHAPTER IX.

1852—1853.

To Charles Bunbury, Esq.

Manchester, *March 1st,* 1852.

MY DEAR CHARLES,—You appear both to have much enjoyed the south shore of the Isle of Wight, and you seem to have found much to interest you in the geology notwithstanding the season is so unfavourable for out-door work.

I suppose there never was collected together in the government of the country so many untried men as Lord Derby has recruited. Lord Derby's speech was good in many parts, excellent in some, especially that in which he spoke so handsomely of Lord Lansdowne; and the moral effect of two political opponents living on such terms of private friendship must be great. But he has left no doubt that he is to try to restore protective duties; and I trust that he will not be long of discovering that the country will not endure it. You will observe too, that he gives very early signs of his intention to favour the Church in its dominant pretensions on the subject of education. I trust that in this too he will be signally defeated. I should be as ready as anyone to give the established clergy the most abundant means and every encouragement to spread education on a sound basis among the children of their own flock, but I must claim from the state the same abundant means and the same encouragement to all other religious bodies. Lord Derby has already roused two feelings that, if I mistake not, will create a ferment, that will either force him to give up protection and the exclusive fostering of the Church in matters of education, or turn him out of his place.

I hope no time will be lost in taking measures for the better defence of the country, for no one would be so foolish, I suppose, as to place the least reliance on the protestations of Louis Napoleon. I have not seen Sir Charles Napier's pamphlet (I mean the General) but the extracts I have read were excellent.

Our affectionate love to our dear Frances,

Yours affectionately,

LEONARD HORNER.

To his Daughter.

Manchester, *March 4th*, 1851.

MY DEAREST KATHARINE,—On the 22nd of last month I had the great happiness to receive your delightful letter from Agra of the 2nd of January, and the same mail brought your letter from Meerut of the 8th. It was that day you wrote exactly six months from the time I parted with you. Your letters are so full of all we desire to know, that we feel quite up to all that is going on around you. . . . . . . Your dear Leonard must be a most charming, engaging creature, and a source of unspeakable happiness to his father and mother. I looked yesterday on the lock of hair you gave me on the 17th January, 1851, and have had in my pocket ever since, with the pretty lines that were drawn at Mildenhall on the last day of the year.

Charles Bunbury has been elected a member of the Athenæum, which will make London far more attractive to him, and will, I hope, lead to their spending more time there. Her book has been successful I think, it is admirably translated and reads quite like an original work, but it requires one to be very familiar with the works of Dante to read it with much interest, which unfortunately is not my case. She has great comfort in her schools, for they are both flourishing, there being eighty-four in attendance at the

girls' school and the last account I had of the boys' school a
month ago gave seventy-two, so that they are now at the
maximum the building will hold, and Sir Henry is going to
enlarge it this summer.

Before this reaches you, you will have heard of the retire-
ment of Lord John Russell and his government, and Lord
Derby being Prime Minister with Disraeli for his Chancellor
of the Exchequer.   It is a most extraordinary Ministry,
scarcely a tried man of business among them.   Lord Derby
having declared that, if he can, he will re-impose a duty on
Corn, has resuscitated the Anti-Corn Law league.   A great
meeting took place here on Tuesday, the day before yesterday,
to re-organize the league ; most determined resolutions for a
general agitation of the question all over the kingdom
were passed, and in twenty-five minutes there were sub-
scriptions for the purpose of carrying the resolutions into
effect, of £27,700!   Fourteen persons subscribed £1,000
each.

I had two days ago a letter from Mr. Harris of Alexandria,
giving me hopes that as soon as the Nile has subsided,
excavations will be made on the site of Memphis, similar to
those made last summer at Heliopolis, which will be very
important, for Memphis was built by Menes, 3,650 years at
least B.C., and if we get below the foundation of any structure
there, and find Nile sediment, it will be very interesting.   I
am waiting with impatience for the engineer's reports and
plans of the work at Heliopolis, which was very extensive.
Mr. Harris sends me a very interesting account of Hekekyan
Bey, who conducted the operations.   He is an Armenian who
was sent at ten years of age to England, where he remained
for twelve years and got a capital education as a civil
engineer.   Being a Roman Catholic, he was educated at
Stoneyhurst College in Lancashire.   On his leaving England
he came to Egypt, was in great favour with Mahomet Ali, and

is so with Abbas Pacha. His assistance has been invaluable, and he enters into the enquiry with much zeal.

God bless you, my dearest Katharine. Give my love to Harry, and manny kisses for me to your darling boy. Remember me to Ayah.

Your very affectionate father,

LEONARD HORNER.

To his Wife.

Manchester, 28th May, 1852.

You ask me to give you a more clear idea than you have of the object of my researches in Egypt, and I will try to do so. According to records that go back to a very remote period, the Nile has every year overflowed its banks at a particular season, to such an extent as to cover with water the low land on each side of it, and that large portion at its *embouchure*, which, from its resemblance to the form of the Greek letter D, Δ is called the Delta, that being the name of that letter. The water during the inundation is very muddy, and as the river has a very slight fall, the mud is deposited on the land over which the water spreads. As the amount of the inundation is on an average of years' uniform, so also is the amount of mud deposited uniform, and it has been calculated, by some very exact observations of the French, at the end of the last century, that every hundred years the mud deposited amounts to five inches. This amount, however, is subject to certain irregularities according to the localities.

If, therefore, perpendicular boring through the mud-deposit be made, for every five inches of that passed through, we may reckon (according to the assumed rate of increase) one hundred years, and thus if five hundred inches were gone through, it would indicate ten thousand years since the lowest of the five hundred was deposited.

These researches have no relation to my former enquiries
into the accuracy of some of Lepsius's observations *in Nubia*.
The spots chosen for the excavations, Heliopolis and Memphis,
are, on account of monuments existing there, the age of which
is known; and it was desirable to ascertain whether there are
regular layers of Nile mud under the foundations of these
monuments.

To his Daughter.

Stirling, *September 16th*, 1852.

MY DEAREST KATHARINE,—Your Mamma with Joanna is
sitting at the same table with me, writing to dearest Mary, of
whose safe arrival we had the happiness to hear yesterday, at
Halifax on the 1st of September.

We passed a most happy week at Bonaly, Lord Cockburn
in excellent health and spirits.

We left Bonaly on Friday the 3rd of this month, went to
the railway station in Edinburgh, and in an hour and twenty
minutes we were in Glasgow. We drove through the Salt
Market and High Street, visited St. Mungo and its beautiful
crypt, and then went by steamer to Greenock, a beautiful sail.
The Tontine at Greenock is a capital inn, and we were feasted
with Loch Fyne herrings. On Saturday the 4th we started
in a steamer up Loch Long and Loch Goil, and at the head
of the latter Loch got into a coach, which conveyed us through
a wild and beautiful pass, which you and your Mamma tra-
versed in 1845, to St. Catherine's on Loch Fyne, where we
found Lord Murray's carriage waiting for us, and in an hour
we were at Strachur. The host and hostess and William
Murray were remarkably well, the latter, who is in his 79th
year, goes out shooting and has no infirmity of any kind.
There were in the house Sir John Hippesley, a Somersetshire
baronet, who remembers seeing your dear uncle at his father's

house, at the time of the Western Circuit, when he himself was a boy; Miss Farquhar, daughter of Sir Walter Farquhar, the banker, whose acquaintance Susan made at the Stewart Mackenzie's in Paris last spring, and liked much, a most charming person whom I greatly admired; Caroline Bell and her brother Frank from Calcutta; Mr. Burton, the author of a "Life of David Hume;" and Mr. Brandling from Northumberland.

On the morning of the 8th we started in a barouche from Strachur Inn, and drove first to a very lovely spot at the head of Loch Fyne, and then over the mountain through Glen Croe to Tarbet on Loch Lomond. We arrived just as the steamer was starting for the lower part of the Loch, and had a fine sail, with Ben Lomond and the other mountains clear to their summits; we dined at Balloch and returned by the steamer to Tarbet where there is a capital inn. Next morning a steamer carried us to the top of the Loch, which I think is the finest part. It brought us back to Inversnaid, and from thence we crossed the mountain to Loch Katrine over a desperate rough road. A steamer on Loch Katrine brought us in an hour to the Trossachs.

Next morning we had a row on the lake, and saw all its beauties in a glorious day. We then drove to Callander in a comfortable barouche, supplied at the inn, and a similar one at Callander conveyed us by the pass of Lenny, Loch Lubnaig, Lochearnhead and Glenogle, to Killin, where we stayed all night, a most beautiful spot. Next morning we started early for Kenmore, saw Lord Breadalbane's noble place of Taymouth and then had a beautiful drive of eighteen miles along the banks of the Tay, by Aberfeldy and Logierait, where the Tummel falls into the Tay, and then up the former river by Moulineam to Bridge of Tilt close by Blair Athol.

On Monday we came to Perth, passing through the grand pass of Killicrankie, and went that day to Alva House (near

Stirling) to visit the Dean of Carlisle, who is occupying that
fine place with his brothers, and had a most agreeable visit.
Mrs. Tait is a charming, sensible person, and the Dean is you
know a great favourite of mine. We had his brother John
with his son and two daughters, his brother James, and his
brother Tom lately returned from India, a brevet Lieutenant
Colonel of irregular cavalry, but a captain only in his own
regiment.

Edinburgh, September 17th. The day before yesterday we
came into Stirling to meet Cockburn, who was to arrive as
Judge of the Circuit Court. He came about three, and we
had a delightful walk with him round the castle, and we afterwards dined with him, the Dean and John Tait being of the
party. We returned to Alva House at night, and came back
again to Stirling yesterday, and were present at the trial of
two prisoners. We thought that Cockburn executed the
duties of judge with great dignity. He had to give an official
dinner to magistrates and other persons of local importance ;
we dined by ourselves, and afterwards drank tea with the Miss
Spiers, our old friends, who live at Laurel Hill, about a mile
from Stirling, a very nice place. We had a very merry evening
afterwards with Cockburn, John Tait and Mr. Handyside, the
sheriff of Stirlingshire. We left Stirling this morning with
Cockburn and got here about 12 o'clock, and we mean to
remain for a week and then go to Manchester. We have seen
Frances and Charles, who are in lodgings in Melville Street,
they hope to make a short Highland tour.

We are looking with some impatience for another letter
from you. I trust none of you have suffered from the great
heat. We were much comforted by hearing from Colonel Tait,
that the climate of Umritsur from October to April, is as cool
and pleasant as autumnal weather in England. This I hope
will brace you all after the relaxation of summer. Your
accounts of all the little ways of your dear boys are read with

the greatest interest, particularly the proofs of intelligence, merriment, and activity of dear Leonard, and you cannot be too minute in your accounts of them.   Harry appears to make most excellent and thoughtful arrangements for the comfort and health of all.

Of public news the most important is the death of the Duke of Wellington. Last Tuesday the 14th, he rose and break-fasted as usual, took a walk, and shortly after had an epileptic fit, similar to one he had several years ago, but this time he had not strength to rally, and he expired in a few hours, a happy release after so long a life. His death forms quite an epoch in our history. There will be much speculation as to who will be Commander in Chief. Lord Hardinge is in the mouths of most people, Prince Albert and the Duke of Cam-bridge are also speculated about. I hope the Prince will not take it if offered, for it would be a great hindrance to his use-fulness in the lines where he has already proved so successful.

And now, my dearest, I must say farewell for the present. Your dear Mamma is upon the whole very well, and we do not allow her to take too much exercise.

God bless you, my darling, my love to your good husband, and kisses to the two dear boys, and best regards to Ayah.*

Your affectionate father,
LEONARD HORNER.

To his Daughter.

Manchester, September 30th, 1852.

MY DEAREST MARY,—It was a great joy to me yesterday to receive your dear letter from Fredericton with Charles's

---

° A most valuable Indian nurse who was with Mrs. H Lyell before she left England.

interesting postcript.   I could not help admiring Charles's
zeal in starting with so little delay with Mr.   Dawson, and I
long to hear further particulars of his discoveries, particularly
as to the bones found in the filling up of a fossil tree.   Your
voyage with the Heads and your visit to Fredericton have been
a very pleasant addition to your contemplated American tour.
I hope you have been long safe and well with your many kind
friends at Boston.   Give my kind regards to Mr. Prescott and
his son, and to Mr. Everett, I wish I knew more of the
excellent people you have, by your description, made me
acquainted with.

We made out our Northern tour most successfully, and had
fine weather all the time.   Our visits to Mrs. Fletcher and
Mrs. Davy were very pleasant, and no less so, that to the
Speddings.   Then   our   week   at   Bonaly   passed   very
satisfactorily, for Cockburn was in excellent spirits, and in
point of health and cheerful activity he is as well as he has
been any time these ten years.   His man, Philip Brodie, the
farm servant, observed to me, "My Lord is as souple as an
eel."   Mrs. Cockburn is very cheerful and looks well, but she
can walk with great difficulty.   We started for the Highlands
on the 3rd, and by Glasgow and Greenock got next day to
Strachur, where we passed four days very pleasantly.   We
got to Stirling on the 14th and paid a visit to the Dean of
Carlisle at Alva House, Mr. Johnstone's, which he and his
brothers had taken for four months.   There was a large, happy
family party.

During our whole stay in Scotland I never heard the words
"Agricultural distress."   Fergus, the member for Fife, told me
of a large farm out of lease last November, which was let at
an increased rent, there being eleven competitors after it;
William Murray let a farm last January in Dumfriesshire, at
an advance of ten per cent on the rent, and for which there
were fourteen competitors, and Sir William Craig had thirteen

competitors for a farm lately, but which he let without raising
the rent.

The Free Kirk seems to be going on with unabated zeal, for
we saw new churches building, one very large one at Stirling,
and the sustentation fund, *that* out of which the stipends are
paid, amounted last year to £123,000. Besides that large
sum, supplemental funds are collected by congregations to
raise the stipends of their ministers, several of whom in the
large towns get as much as £500 a year. But I did not learn
any *good* that the disruption, as it is called, has yet done ; and
when I asked Cockburn if he could tell me of any good, he said
"none, but much mischief." This surprised me, for he had
always upheld to me the justice of the cause of the Free Kirk.

I have just finished reading "Uncle Tom's Cabin," and
could scarcely lay it down. It is most painfully interesting.
I cannot think so badly of human nature as not to believe
that it will make a great impression on many who have
hitherto thought little of the true state of the slave question,
its horrors, which can in no way be salved over by the
comfortable condition even of most slaves, *as* animals, as
chattels. Without going the length of abolition, except by
very gradual steps, both for the sake of the owners and the
slaves themselves, there surely might be laws passed by
congress to prevent many of the cruelties—especially the
separation of parents and children—and the prohibition of
corporal punishment. One is too much habituated to associate
slavery with negroes only, forgetting that in its worst features
it affects beings who have only a small portion of negro blood
in their veins. I doubt whether the worst conduct of the
slave-owner in the Southern States is more cruel than the
proscription of the Mulatto, and black blood, however diluted,
in the Northern, for the mental suffering of the latter must
be often more difficult to bear than the worst evils the poor
ignorant negro is subjected to. I should like to hear what

impression has been made in the United States, and especially
in the slave states by this book.

Ever my dearest Mary's affectionate father,

LEONARD HORNER.

To Sir Charles Lyell.

London, *18th November*, 1852.

MY DEAR CHARLES,—We have just come in from the
Athenæum, from which we saw the procession of the Great
Duke's funeral. I got my ladies safely deposited in the
gallery erected outside, nearly at the angle of Pall Mall, where
they had a most excellent view. Susan and Joanna received
tickets for St. Paul's from the Dean, and they made an
arrangement with Lady Herschel that their carriage was to
follow hers. They started at five in the morning and now at
nine o'clock this evening they are just returned safe and well.
The procession that we saw was most striking, particularly
the numerous regiments of fine troops of all kinds. It started
from the Horse Guards at eight, up Constitution Hill, passed
Apsley House, and St. James' Street, Pall Mall, and so on.
It reached the Athenæum at half-past nine and was an hour
and three-quarters in passing. There was an enormous
crowd, and nothing was more striking than the entire silence
of the mass, and when the bier of the poor Duke passed the
end of Waterloo Place, they all took off their hats. After
many days of rain and a threatening morning it cleared up at
eight o'clock, and during the whole procession there was
bright sunshine. By Susan's and Joanna's report of what
took place in St. Paul's it must have been a grand sight,
although more of theatrical display than was consistent with
the solemnity of the occasion. The day has gone off well and
we have heard of no accidents.

Give my love to my dearest Mary. May this find you both

well, and may you have a safe and good passage across the
Atlantic and joy us all by your appearance about the 13th or
14th of next month.

Yours affectionately,

LEONARD HORNER.

To his Daughter.

Highgate, 23rd January, 1853.

MY DEAREST KATHARINE,—You will readily believe what
pleasure your letter gave us, as it is written in such good
spirits, and gives so good an account of yourself and Harry,
and the dear boys. I am relieved that you have taken a house
at Simla, and I trust you will get well established there, before
any oppressive heat commences.

Nora and Joanna are going on most diligently with their
translation of Lepsius, and I think they are doing it most
excellently. They are getting assistance in different passages
from the Chevalier Bunsen, and Dr. Hofmann. The latter
was at the Barlow's the night before last, after Faraday's
lecture, and he introduced me to a son of Liebig's. Liebig
has left Giessen, having got a better appointment at Munich.
Faraday's lecture was on magnetic force, it was not adapted
to a popular audience, but there was a part of his lecture that
was intelligible and interesting to all He described the
remarkable observations of Schwabe of Berne, on the spots of
the Sun. He has been observing them diligently for many
years, and finds that while at one period the Sun is quite free
from spots, they gradually increase until they reach their
maximum at the end of five years, when he counted as many
as 333 *groups* of spots. From that time they begin to decrease
in number, and at the end of another period of five years, the
Sun is again free from spots. This series he has observed
through two periods of ten years. But the most interesting

fact I have got to notice. It has been observed that the variation of the magnetic needle increases and diminishes also in a period of ten years, and that the greatest amount of variation takes place when there is the greatest number of spots on the Sun. These facts were mentioned to shew, in addition to other phenomena, that there exists some magnetic relations between the Sun and our earth.

Charles Lyell's paper last Wednesday at the Geological Society on the discovery he made in Nova Scotia, on the Bay of Fundy, of reptilian bones, and a land shell in the loose materials filling up the hollowed trunk of an upright stem of a sigillaria, in a bed of coal, excited a great deal of interest, as supplying the first instance of a *land* shell in a bed of coal belonging to the old coal formation, and another (the first in America) of the existence of animals of that advanced organization at so early a period.

I received a week ago from Hekekyan Bey, a report of the operations he carried on for me last summer on the site of ancient Memphis, opposite to Cairo, a little higher up. They have obtained many remarkable results. The details of the sinking of thirty-three wells on a line between the Libyan and Arabian ranges of hills I have not yet got, but he describes some discoveries of wonderful colossal Statues, two of granite, twenty-five cubits in height, more than forty feet. Mr. Birch of the British Museum came last Tuesday to see the interesting specimens of idols and objects of pottery dug out at great depths which he has sent me, and I am going to-morrow to Bunsen's to shew him the report and some drawings. Mr. Charles Augustus Murray is now in London, but sets out on the 4th to resume his place of Consul-General in Egypt, and most kindly offers his valuable aid in the prosecution of these researches. Abbas Pacha, the Viceroy, has not only been liberal but munificent in the way he has had all these extensive operations carried on, at his own expense. It

will be a long time before I have all my materials collected for
my projected essay.

Bunsen's recent work, " On Hippolytus and his age," is
calling forth great excitement. Hippolytus was a bishop at
Rome in the second century A.C., and a work of his, discovered
some years ago at Mount Athos in a monastery, is that which
Bunsen has published with a commentary. It is curious, but
hard reading. I asked Milman what are the chief matters
that are new and important, and his reply was :—" the work
cuts up Popery by the roots."

Bunsen is now working at the continuation of his *Egypter
Stelle in der Weltgeschichte,* and hopes to publish the completion of the work in July. He promises, publicly, a history
and comparison of the four Gospels, and a Life of Christ. He
is a person of the most wonderful industry. He told me that
the recent discoveries of Colonel Rawlinson in Assyria, threw
a flood of light upon some parts of chronology hitherto involved
in the greatest obscurity.

God bless you, my very dear Kate. My love to Harry and
many kisses to my dear grandsons.

Your affectionate father,

LEONARD HORNER.

To Major Lyell.

Manchester, *March 17th,* 1853.

MY DEAR HARRY,—I have to thank you for the " Friend of
India " of the 6th of January It contains a great deal that
is interesting, particularly that sketch of the gradual and
rapid extension of our Empire in the East. You will see that
the affairs of India are beginning to occupy a great deal of the
attention of Parliament, connected with the great question of
the renewal of the Charter. That there will be some considerable changes in the power and management of the

company, in respect to the details of Government, is very clear. One thing is evidently to be changed, viz., that young men, or rather boys, sent out as writers, will no longer be allowed to be appointed as magistrates and judges. It has always appeared to me a mockery, and a gross injustice to the natives. When one considers the enormous extent of the territory, and the wide population, the whole government is a matter of wonder.

I trust that this letter will find you all at Simla, enjoying the fine climate ; and that we shall have continued accounts all through the summer of the good effects of it. That the darling boys will thrive, and that dearest Kate will not suffer from the heat as she did last summer. I hope too that you will not be much away from her.

Farewell, my dear Harry. Give my affectionate love to my own dear Kate, and kisses from Grandpapa to the darling boys.

Your affectionate father-in-law,

LEONARD HORNER.

To his Daughter.

Highgate, *April 9th*, 1853.

MY DEAREST FRANCES,—I have just been, by Lord Palmerston's desire, preparing a draft of a Bill to amend the existing Factory Acts, which require it in several particulars ; but whether he will adopt my remedy I do not know. I saw Lord Shaftesbury yesterday, and he approved much of my proposals and said he should speak to Lord Palmerston to-day.

We are to have Dr. Hofmann, and Dr. Becker, Prince Albert's librarian, to dine with us to-morrow.

Hofmann has been gaining great repute for his lectures in the Laboratory of the Royal Institution, which are attended regularly by Faraday, Graham, Henry and others of the most eminent chemical philosophers in London. He has been appointed chemist in the Museum of Practical Geology in the room of Lyon Playfair, who has been advanced to the office of Secretary to the new department of Science erected in connection with the Board of Trade, with £1,000 a year. A most excellent appointment to a most useful new public department, from which I augur much good, especially in the spreading of schools in the provinces for practical science. Already he has had a communication with me about bringing the School of Arts in connection with this department.

I hope you and Charles liked Lord John's speech on the education question—I did very much. It will be a considerable step in advance, especially the introduction of a school rate. There are many things I should have liked to have seen done besides, and I daresay if he had had a fair open field he would have proposed much more; but he has shewn practical wisdom in not proposing to do more than he has a reasonable chance of carrying. If as wide a step be made every year or two, we shall have a vastly improved system in a few years. All join in affectionate love to yourself and Charles, with, my dearest Fan,

Your affectionate father,

LEONARD HORNER.

To his Daughter.

Putney Villa, *June* 30*th*, 1853.

Two days ago, my dearest Katharine, I received your most agreeable letter of the 10th of May, and rejoiced I was to learn that Harry was with you and well, and that the dear boys were also well. I hope that they are both of a constitution to thrive in the keen air of Simla. It must be very beneficial to Harry to have been removed from the broiling heat of Umritsur to the cool region, and I hope it will in no degree interfere with any of his *regimental* interests.

The question of war between Russia and Turkey is not yet settled, and if it should take place it will be war also between Russia and France and England; and how many other powers consider it for the interest of the civilized world that Turkey should not be replaced by Russia, and will therefore lend their aid to Turkey, one cannot tell. I cannot believe that the Czar will be so mad as to attack Turkey, and therefore believe that peace will be maintained.

You are aware that Robert Stephenson, the engineer, is engaged in making the railway from Alexandria to Cairo. Thinking that I could obtain some useful information from him relating to my Egyptian researches, I sought to be introduced to him. I have had two long conversations with him, and he has given me the most valuable assistance relative to the levels and geological structure of the Isthmus of Suez, including maps and specimens. It is now clearly made out that there is no difference between the sea level at low water of the Mediterranean and the Red Sea. It was for half a century supposed, on the authority of measurements made by the French engineers of Buonaparte's expedition, that the Red Sea was thirty feet higher than the Mediterranean. I am in daily expectation of receiving three boxes sent off by Hekekyan Bey, containing sections of specimens relating to his operations last summer at Memphis, but I shall not be able to open them for five weeks at least, as on Tuesday next I go to Lancashire. After I have done my work I mean to go for ten days to Edinburgh. I shall be there during the Exhibition of the Academy, and shall probably spend my time between Bonaly and the Murray's in Great Stuart Street. The Academy continues to prosper, and so does the School of Arts. The statue of Watt, which I induced the Directors to agree to having, will, I understand, be ready to be erected in

January. It is a copy of that by Chantrey which is in the church of Handsworth near Birmingham, where Watt was buried, and is being executed by a rising artist in Edinburgh, of the name of Slater, in sandstone obtained near Granton, and the Duke of Buccleuch gave the block. It is to be placed in Adam's Square in front of the school, and will be a great ornament to that place, and will be seen by every student of the University as he passes daily to the College.

We came here two days ago.* It is a beautiful place by situation, and noble trees, and well laid out. Mr. Moore is as well as I have seen him for several years, and his mind is as alive and active as ever. They are most hospitable and kind.

I am ever, your affectionate father,

LEONARD HORNER.

To his Daughter.

Skipton, *July 12th*, 1853.

MY DEAREST MARY,—You will readily believe that I have been thinking much of you and dear Charles on this day, a day I have had so many good reasons to bless. May I never see it without the power of knowing that you are both well, happy and prosperous. I am most thankful that you have felt no bad effects from the great heat at Washington, and it is extraordinary that it gave Charles no annoyance. What a mercy it was too that you escaped unhurt in that frightful thunderstorm so near you. You seemed to have passed some very interesting days at Washington, and Charles's interview with the President must have been particularly so. What an advantage to a country it must be when the Chief can communicate freely with all capable of giving him information;

* Staying with Mr. Carrick Moore.

an advantage which the etiquette of European Courts so much stands in the way of. I am glad to hear that the Exhibition is certainly to open on the 15th, and I hope you will arrive at New York to-morrow after successfully accomplishing all you projected when you wrote on the 28th of June. I trust too that you will be able to embark for England as early as you say. Unless I am sent for by Lord Palmerston before, which is just possible, because of a new Factory Bill that has been brought forward by the member for Oldham, Cobbett, I hope to be at home again about the 10th of August, ready to receive you.

Three days before I left home I received some valuable communications from Cairo, the account of the operations last year at Memphis, which I long to be at work upon. There are several plans and maps, executed by Arab Engineers, and most beautifully, and with every appearance of correctness.

Mr. Murray sent me from Linant Bey, the chief engineer in the Pacha's service, a series of tidal observations at Suez, and the ancient Pelusium from which it has been established, as Robert Stephenson had done before by actual survey, that there is scarcely any perceptible difference between low water mark in the Red Sea and Mediterranean. Stephenson is of opinion that there are clear indications of movements of the land in recent times, even within the historical period. He has given me many specimens, among them several organic remains, which I have had examined by Edward Forbes. Forbes said, "Why does not Lyell go to Egypt? it is just the country for him, an almost unexplored field, and very easily examined, far more in his way than the Canary Islands." Now this was quite spontaneous on Forbes' part, no prompting by me.

I have during this last week visited two admirable British schools, one in a retired valley, and I was astonished at the variety and accuracy of the knowledge (secular) which the

children had acquired, and with a fair knowledge of the
historical part of the Bible.

My love to Charles,

Your affectionate father,

LEONARD HORNER.

To his Wife.

Newcastle, *20th July*, 1853.

I for the first time enjoyed the benefit of the electric telegraph to-day. I left my umbrella at a station near Durham,
and when I got to Newcastle I mentioned this to the station
master there, he sent a message to the place, and in three
minutes an answer was returned, that the umbrella was found,
another message for them to send it to Newcastle by the first
train, for all this they charged *nothing*. The umbrella is
again in my possession. I have been having a correspondence
with Dr. Lyon Playfair respecting a scheme of introducing
some instruction in physical science at the Academy, and he
agreed to meet me in Edinburgh both for that and for a
meeting with the directors of the School of Arts to bring it in
alliance with the new Government department of Science.
He wrote to me that he should be in Edinburgh this evening,
so I waited for the arrival of the express train at five, found
him in it, and we settled our plans of meeting. He is to dine
at Bonaly on Saturday, where Cockburn has invited the Rector,
three Academy masters, and three directors.

To his Wife.

Bonaly, *July 24th*, 1853.

After breakfast on Friday, I called first on William Playfair,
to tell him how much I was pleased with his new building on
the mound. It is now so far advanced that one can form a
very good idea of what it is to be. It is in my opinion tho

most chaste, beautiful structure in Edinburgh, and I do not
know any modern one elsewhere that pleases me so much,
Edinburgh is much indebted to Playfair for having converted
the hideous mound into a great ornament of the city, besides
the many other architectural adornments erected by him.
He is, I am sorry to say, very feeble, and can scarcely walk.
I had a long talk with the worthy Dr. Murray, our indefatigable secretary, and then drove to Brown's Square to have
some conversation with our lecturer on chemistry, a very
able and excellent man, in reference to the plan suggested by
Lyon Playfair, of an alliance between the School of Arts and
the Government new department of science in the Board of
Trade, of which Playfair is secretary. I then proceeded to
Great Stuart Street, and after luncheon, Murray and I drove
to a sculptor's of the name of Brodie, who is doing a bust on
a small scale of Cockburn, about the size of that we have of
Oersted, and it promises to be like. I am to have a cast of it
when finished. We then called on Pillans, who was out, and
walked first in the beautiful Botanic Garden, and afterwards
in that of the Horticultural Society. There is the largest
class of botany this year that has been known, no less than
232, and Professor Balfour started on Thursday with his class,
for a two days' excursion in the Cumberland mountains.

Yesterday, after breakfast, Murray took me to Steel's studio.
He is getting well on with Jeffrey's statue, which promises to
be good, and he has been successful in a bust of Lord
Mackenzie. He shewed us a design (a model) for the Manchester statue of the Duke of Wellington, for which he among
many others is going to compete, and which I think very fine
indeed ; I like it even more than that in the front of the
Register House. At three I came out here, and had a short
walk on the hills before dinner. I have not seen Cockburn
looking better for many years.

July 28*th*. On Tuesday I went to the Academy early, and

stayed till half past two.   Afterwards, before going out to
Bonaly with Cockburn in the gig, I called on the Murrays
who had returned from paying the Cunninghames a visit at
Duloch.   They were much afflicted by the death of Mr.
Schwabe, which they had just heard of.   Count Flahault was
in Murray's room, sadly changed from what he was, so old-
looking.

Yesterday Cockburn and I started early, and were in Barry's
Hotel at nine, to breakfast with the Academy Club, the former
pupils.   We had a large party, including the rector, masters,
and the duxes of all the classes   At twelve the exhibition took
place, a most interesting sight as you know.   The head boy,
Luke, a very remarkable boy, who besides the gold medal
gained many prizes, is the son of a small baker near Stock-
bridge, and it was very pleasant to see how warmly he was
greeted by the whole school.   The prizes were given by Mr.
Arkley, a lawyer, who twenty years ago had been himself dux,
and the Dean of Carlisle sat by his side.   After the exhibition
some of us walked to Great King Street, to Dr. Hannah's, the
rector's, house, where we had a very pretty luncheon, and
then Cockburn, Playfair, and I, drove round Arthur's Seat,
and by the Grange, paying a short visit to Mr. Maclaren, the
geologist, who bought Dr. Thomson's house of Morland.   We
then went at six o'clock to the directors' dinner, which went
off very well indeed, and Cockburn and I got back here,
between eleven and twelve.   My health was drank with very
kind expressions of good will.

***

To his Daughter.

Highgate, 18*th August*, 1853.

MY DEAREST, DEAREST KATE,—I am sitting in our pretty
drawing-room after dinner, and Joanna has just been playing
to me on her organ " Wilt thou be my dearie," and " Gin

living worth could win my heart," and Nora is working hard
at the Index to their Lepsius, Mamma helping to paste the
slips, and there is your dear picture, and I fancy you singing
"dinna forget." It is a calm evening and there is a fine
sunset, and Polly is talking away in the garden at a great
rate. But it is far from August weather, as it should be, for
we have a fire and find it very comfortable. It has been the
worst year for weather I ever remember. I do not think that
all the good sunshiny days we have had since the first of
January would amount to thirty. I have not given a full
account of our family circle, that is, our more limited Highgate
one, by not telling you that Susan went yesterday to Mildenhall, and she gives a good account of dear Fan and Charles.
Charles and Mary dined with us the day before yesterday.
We had Charles Augustus Murray, Babbage, and Mr. Clough,
Blanche Smith's betrothed. Murray lately arrived from
Cairo, and is appointed Ambassador to the Swiss Confederation.
His successor as Consul-General in Egypt is Frederick Bruce,
a brother of Lord Elgin.

We go to Waverley next Tuesday, the 23rd, and shall
probably stay ten days.

Give my love to dear Harry, and kiss the darling boys
twenty times from me.

God bless you, my own dear Katharine,

Your loving father,

LEONARD HORNER.

_____

To his Daughter.

Manchester, *September 27th*, 1853.

MY DEAREST KATHARINE,—We are most thankful that you
are all at Simla, for the heat of this summer in India, we are
told, has been unusually great. Your letters were particularly
delightful to me, because there was so much about your

darling boys, every little incident respecting them I read with
the greatest interest.   God bless and protect them ; and if I
am so blessed as to see them and have them about me, it will
be the greatest joy of my remaining life.   Your letters are
delightful.

Your Mamma is upon the whole well, more easily fatigued
by walking than she used to be, but we cannot control her
spirit, which makes her do more than she ought.   We went
on the 20th to Mildenhall, and stayed there till yesterday, and
a very happy week we had.   Our dearest Frances is decidedly
better ; Charles had a little cold, but was otherwise well and
cheerful.   I love him more and more, every time I see him.
I do not think there exists in the whole world a more pure
and lofty mind, or a more kind, gentle, unselfish nature.
Then he is so fair and candid, and so accomplished, and he is
a most devotedly kind husband.   They are perfectly happy in
each other.

The proposal of Charles and Mary and Leonora going to
the Canaries has led Frances and her husband to think of a
similar trip.   If these schemes are carried out, your Mamma
and I will be deprived for four or five months of all our
chickens except Susan and Joanna, of whom we must make a
good deal, as I am sure they will do of us.   We may perhaps
make a run of a fortnight or three weeks at Christmas to
visit your aunts (in Paris), but this is as yet a mere chance.
Leonora is very well, working hard at Spanish.   The industry
and perseverance of her and Joanna are above all praise.

29th September.—We are in a great state of uncertainty as
to whether we shall have war in the East or not, and if war
in the East, then war between England and Russia.   Matters
have gone so far by the unprincipled conduct of Russia that I
do not see how war can be avoided, consistently with honour.
That seems to be the prevalent feeling, and that ministers
have acted an unjustifiable part.

The funds have fallen immensely. Joanna has had a capital letter from Lord Cockburn congratulating her and Leonora on their translation of Lepsius. He gives a good account of all at Bonaly. Poor Sir William Hamilton had a fall and broke his arm; we have not heard how he is since. Mamma has written to Lady Hamilton.

God bless you, my dearest Kate, and your dear husband and your darling boys. Remember me to Ayah, what a treasure she is, I hope we shall see her again in this country.

Your affectionate father,

LEONARD HORNER.

From Dr. Tait, Dean of Carlisle.

9th November, 1853.

MY DEAR MR. HORNER,—I received your letter just as I was on the point of leaving home, and have put off answering it till I could command time. It was a most curious coincidence your meeting with my old teacher; I was eight years old when at Whitworth with my brother Campbell, who has now been dead thirty-four years. I cannot tell you how I was moved by being carried back to those old days, when Betty was in charge of her two lame boys under the Whitworth doctors. It is very kind of the old man to have kept so lively a remembrance of me, and I hope if you see him again that you will not fail to tell him how much I value his remembrance of me. It must have been thirty years after he taught me, that he first saw my name afterwards, when gazetted for the Deanery, and it is very curious that he should have remembered me. I shall write to-night to Dr. Cotton, the Provost of Worcester College, whom I know very well, to ask him about his son. Meanwhile, I should like to send the old man a present of a volume of Sermons I published when I left Rugby, if you could let me know his address at Rochdale.

It has been a great pleasure with me, mixed with other feelings, to have the remembrance of these days renewed. I suspect the old man must have taught me writing, and I do not know that he has any great credit in my perfomrances in that way. It was a very strange life we led amongst the rough Whitworth doctors, and their equally rough patients ; but I believe it was the providential means of enabling me to pass through life without being a cripple.

I am very sorry we have missed seeing you here this season ; but trust we shall meet before very long.

Two days ago I had a letter from my friend the Curate of Keswick, reminding me that I had promised to ask you about the Bobbin Mills : whether they are under inspection, and whether there are restrictions as to age. Some frightful accidents have happened from the employment of very young people.

We had a grand Educational Reunion here last night, of which I think I shall be tempted to send you the account when it appears in the local paper.

I have been much distressed within the last few days by hearing of the proceedings against Mr. Frederick Maurice at King's College. I trust it may not be the commencement of an alarming struggle in the Church of England.

Pray give our united kindest regards to Mrs. Horner and your daughters. Do not forget, if you can, to send me the direction at Rochdale.

Believe me,

Yours very sincerely,

A. C. TAIT.

CHAPTER X.
1854.

To Sir Charles Lyell.

Athenæum, *February 22nd,* 1854.

MY DEAR CHARLES,—I hope this will find you all safe and
well at Teneriffe, and that you will be able to accomplish there
a great deal of what you projected.* . . . .
The anniversary of the Geological Society went off very
satisfactorily. I think I wrote to you that we had given the
Wollaston medal to Griffith for his map of Ireland. I believe
there are few geological maps better executed or containing so
much information. Forbes made a very good address to him,
shewing how he had been an active promoter of geology in
the field, for more than forty years.

We never had a more interesting or valuable address than
Forbes has given us. His eloge of Von Buch was very good.
As you may suppose, his main subject was the higher philosophy of the distribution of organic remains, and you will, I
am sure, read it with interest and profit. Hooker's journal of
his travels in India is just out, and he has made me a present
of it. I have read about half of the first volume, and have
found much that is valuable and interesting in it. Hooker,
his wife, and Dr. Thomson are coming to take a two o'clock
dinner with us on Saturday and go to the Zoological Gardens
afterwards. I take great pleasure in going there, Mamma
and I make it our usual walk,† they have just got a second

° Sir Charles and Lady Lyell and Mr. and Mrs. Bunbury were visiting
Madeira and the Canary Islands.

† Mr. Horner had moved his residence to the neighbourhood of the
Regent's Park.

ant-eater, and a female hippopotamus is arrived at Cairo, on its way to our garden.

But you will perhaps wonder how I can write on any other subject than that which is uppermost in every man's mind throughout Europe, the question of war, which is now hardly a question, but as near as possible to a certainty. Louis Napoleon sent an autograph letter to the Czar with an opening for peace, accompanied by a due share of *soft solder*, which Nicholas has replied to in anything but a pacific strain. All hopes of peace seem now to rest upon the possible contingency that Nicholas may meet with the fate of his father. The most active preparations for war, by sea and land, are in progress here. Many of the great steamers are taken up by Government to convey troops, several regiments are already arrived at Portsmouth, Southampton, and Plymouth, and some will sail to-morrow for Malta, 12,000 troops are to be sent immediately, and 12,000 more a month hence. A very powerful fleet is ready to sail for the Baltic as soon as the waters in that sea are navigable. There is an almost universal feeling on the part of the mass of the people for war against Russia, and recruits are coming forward with the greatest alacrity.

Ever affectionately yours,

LEONARD HORNER.

To his Daughter.

17, Queen's Road West, London, 3rd *March*, 1854.

MY DEAREST KATHARINE,—In that delightful letter which announced your intended departure from Umritsur, you say that you had been disappointed by not having had a letter direct from me by the two last mails. I was not aware that I had ever missed more than one, and even that was wrong, but the sole reason was, that you were getting by the same

opportunity so many letters from your Mamma and sisters
that I considered they had left me nothing to tell. But it
was very wrong to run the risk of my dearest Katharine being
disappointed in the slightest degree, when she is so far away.
I fervently hope that in little more than a month I may be
blessed with the sight of you, and that we shall never again
be very far asunder, beyond a very short time.

It will be such a delight if we should see our scattered
children all around us on the occasion of dearest Leonora's
marriage.* It is impossible not to think with sorrow on her
leaving her own home, but that is an accompaniment of every
marriage, and very often with a prospect of a far wider
separation. 1 have the strongest conviction that it is an event
which will add to dearest Nora's happiness. He is so truly
excellent a man, from his childhood upwards he has been a
favourite with all, so Mrs. Brandis says in a letter your
Mamma had from her yesterday. They were much together
as children, and went to the same church in Hanover.
Bernard Brandis dined with us yesterday, and returns to
Bonn on Monday.

You will find this country and France at the commencement
of a war, which I trust will not be of long duration, and end
in the discomfiture and punishment of the disturber of the
peace of the world. I suppose no war ever was undertaken
with a stronger feeling of its necessity and justice, and there
is an almost unanimous feeling in the country that it has been
so entered upon. But the sword once drawn, no one can
predict when or under what circumstances it will be sheathed.
We must make up our minds to a considerable increase of the
Income Tax, so long as it lasts. Is it not singular that my
two schoolfellows, James Deans and Charles Napier, both of
whom I knew intimately, and who were themselves intimate

---

* Engaged to Chevalier Pertz, Royal Librarian at Berlin.

friends, should command the two great fleets that England is sending forth, one to the Black Sea, the other to the Baltic.

God bless you, darling, my love to your good husband, and kisses to the dear children.

Your affectionate father,

LEONARD HORNER.

To Sir Charles Lyell.

17, Queen's Road West, *March 7th,* 1854.

MY DEAR CHARLES,—I trust that this will find you all safe returned from the Grand Canary, busily engaged in the volcanic intricacies of Teneriffe, and much interested in them. We were, you may well suppose, much delighted by the budget from the Lazaretto, that you had had so good a passage from Madeira, and had not found the confinement so irksome as you expected. I was a gainer by it, for you had time to write that very interesting account of your researches in Madeira. It has been read by Forbes with much pleasure, and I sent it this morning, together with Bunbury's very interesting letter, to Darwin. What materials you appear to have already collected for a valuable memoir! and Bunbury too, will be able to give the botanical, as well as the geological world, much information. I do not like the idea of what you write being confined to the very limited sale of our quarterly journal, I think that you and he might together make a book that would have an extensive sale. But I suppose you will find your time far too short to enable you to examine in sufficient detail the various phenomena that Teneriffe presents, and will be tempted to give another season to it, together with Palma and Gomera. There has been no account the least to compare with that which you will be able to give of

Madeira, and it appears to afford most valuable facts for the developments of volcanic action—very puzzling certainly. You talk of reconciling the *Erhebung* system to a certain extent with your former views, but does that go farther than this, that after certain out-pourings of lava have taken place the solid rocks so formed had, by the same volcanic force been heaved up to a higher angle in consequence of the ordinary orifice being stopped by some means? This is not *Von Buch's Erhebung* theory, which supposes a series of rocks poured out on a level surface, and then raised up with a qua-qua versal dip. As far as I can understand his account of the Peak of Teneriffe, or rather of the whole island, I see nothing to support that theory. I hope you will be able to investigate very thoroughly whether there be any good foundation for his asserting that there are two distinct series of volcanic products in the island, marked by mineralogical differences, or whether there be not an identity in many instances of the products of eruptions, the date of which is known, and those of what he would term the materials raised up by his theory of elevation, prior to the pourings out in historical time. Great difference may always be expected in subaqueous and subaerial lavas, but are not the lavas of subaerial volcanos very frequently undistinguishable from those that have flowed out under the sea?

You will readily believe the joy with which we look forward to the prospect of seing dearest Kate and her husband and three boys before the expiration of another month.

War will almost certainly now be formally declared in a fortnight or less, for Nicholas has just issued a manifesto that removes all reasonable expectation of his coming to his senses. By this time a considerable number of our troops have arrived at Malta, and at this moment they are giving a banquet to my old friend Charlie, who is to command the great fleet that is to be sent to the Baltic as soon as the ice is gone.   Gladstone brought forward his finance statement last night.   The Income Tax is to be doubled the next half year, and that will probably last for a considerable time.

My dear Charles,
Yours affectionately,
LEONARD HORNER.

From Mr. Edward Forbes.

College Museum, Edinburgh, *May 24th*, 1854.

MY DEAR MR. HORNER,—The exceeding pressure of business here has prevented my writing sooner, I find such lots of things to do, and so much to be looked over in the way of preparation, called, as I have been, to a lecture at a moment's notice.   I am very glad you approve of what I said at the beginning, and I have every reason to be gratified with my reception here.

The stock of material for a natural history course is indeed very surprising in this Museum.   Jameson seems to have taken great pains to get together the best illustrations of everything that he *did not teach*.   Thus I found as complete a *teaching* series of fossils, arranged in perfect order *throughout the formations*, as could be wished, made to order by Krantz of Dublin and filled up from other sources.   It is the same with geology.   There is much here to satisfy me that Jameson was a very remarkable man—though a very odd one.

I have above a hundred students, far more than I expected, for until three days before I commenced, my coming was uncertain, so that the students were obliged to complete their arrangements in order to register before I came.   My class holds one hundred and sixty people, and is daily full, since many come through curiosity or love of science.   I see many persons there now that the Assemblies are meeting.   Balfour has two hundred and ten students, and I might have had as

many, could the class have begun as usual.   Pillans is as well
and boy-like as ever, and we shall be good friends.

Give my best regards to Mrs. Horner and all the ladies and
friends about you, I shall write more at length when I get a
little more leisure.

Ever, dear Mr. Horner, with many thanks for your kind
services, very much yours.

EDWARD FORBES.

To his Wife.

Home Office, 20*th June*, 1854.

On Saturday I went to the Grove's alone ; a large evening
party, several whom I knew, but I talked much to Mrs.
Bonamy Price and to Wheatstone.   Mrs. Price expressed some
excellent opinions on the respectability of those women who,
left in poor circumstances, do something for themselves rather
than live on the charity of relations.

On Sunday dear little Leonard breakfasted with us and
was as good as possible.   I have proposed, and Kate and
Harry have consented, that Leonard and Frank should breakfast with us every Sunday.   It will be something for the dear
pets to look forward to, and perhaps something to look back
upon with pleasure in their after life.   Goethe did so at his
grandfather's, and speaks of it in his autobiography with
pleasure.

I shall be delighted to have you back on the 30th.   Nothing
reconciles me to my solitary room, except the hope and belief
that your stay at Mildenhall will do you good.

God bless you *dearest, dearest*.   Love to Fan and Charles.

From Edward Forbes.

College Museum, Edinburgh, 26*th June*, 1854.

MY DEAR MR. HORNER,—I should have written before but
the preparations of my lectures keeps me in perpetual move-

ment, and the getting the museum into preparation, for getting it into proper shape, leaves little time to spare, over and above occasional hospitalities.

I have every reason to be pleased with my class, both as to number and quality. The number now is 147 ; some of these however, are Jameson's perpetuals. The actual class is more, because I have always a number of visitors, and several of the Edinburgh doctors of my own standing, or older, are regular attendants, as well as sundry *Savans* and *literati*. I am very glad to see them come, and shall never throw difficulties in the way of entrance into the class-room door. A liberal policy in this matter is the best.

I have taken to the fields on Saturdays, with great success, and the demand for hammers has risen exceedingly. Last Saturday's lecture was given in the King's Park. Maclaren has most kindly been acting as my mentor in the geology of the neighbourhood, and Fleming is very hearty too. There is a great amount of unroused talent here, if it could be got into harmonious working, and I don't see why it should not be so. By the institution of *Red Lionists*, I hope it may be done, and there are some capital and clever fellows to be found at hand and ready. We have even a nucleus of old Red Lions to begin with, for Balfour, Bennett, George Wilson, Blackie, Captain James, and Chambers, and all such. My name-sake has just returned to Edinburgh, much better, but looking still very ill. He came to me as soon as he arrived. There is good news too of Gordon, who is at Carlsbad. On the whole I think the University here prosperous, and the chairs mostly well filled—the medical ones excellent. The college divinity professors, too, are apparently strong-headed and liberal-minded men, above the ordinary run of pious prejudices. Aytoun is the wit of the University, and very clever and amusing he is. Pillans is the patriarch, but as lively as a boy.

I shall be very much obliged to you if you will spread the
Addresses in Germany. It is very pleasant to hear such good
news of Madame Pertz.

Give my very best regards to Mrs. Horner and all at home,
and believe me, dear Mr. Horner,

Ever most sincerely yours,

EDWARD FORBES.

To his Daughter.

Berlin, 17th *July*, 1854.

MY DEAREST KATHARINE,—We have made a successful
journey hither, first a good passage of the sea, and an entertaining journey to Bonn, and five days most agreeably spent
there among so many old friends, and lastly our rapid progress
to this place with two interesting visits at Braunschweig and
Magdeburg.

Yesterday it was intensely hot, and it appears to be equally
so to-day, so much so that I have not yet gone out, and it is
twelve o'clock. Since breakfast George* has taken me through
the various rooms of the library. It contains about 600,000
volumes, and is a model of order and accessibility, under a
system established by George himself, who seems to be a
heaven-born librarian. It is upon the most free liberal footing,
and the advantages to any man engaged in any department
of study are immense, for the books are lent out. Dearest
Nora looks very happy, and she has made every arrangement
for our comfort.

I read with pleasure all you tell me about darling Leonard
and Frank, but you say nothing of Arthur, and I conclude that
he was well. Remember what I said when I wrote to Umritsur,

* Dr. Pertz.

that on no subject would you dwell upon more satisfactorily to me, than in narrating all the sayings and doings of the dear children.

Your affectionate father,

LEONARD HORNER.

To Sir Charles Lyell.

Berlin, *21st July*, 1854.

MY DEAR CHARLES,—I have renewed my acquaintance with Mitscherlich, Heinrich Rose, Ehrenberg, and Von Decken, and I hope to see Humboldt to-morrow. I have made the acquaintance of Gustave Rose, Ewald and Beyrich.

I went with Hermann Pertz to a lecture of Gustave Rose, whose course he is attending ; it was upon Metamorphism and cleavage planes, and I think he must have shot far above the heads of his students. He, after the lecture, shewed me the many rooms containing the collections of minerals and rocks, which are very extensive, and I saw many very remarkable instances of metamorphism of minerals, such as calcareous spar converted into silex, &c., often called pseudomorphic crystals, but the theory of their formation certainly belongs to that most curious and difficult subject Metamorphism, as yet but very imperfectly understood. The palæontological part of the collection I hope to be shewn by Beyrich. I sat half-an-hour with Ewald, who sets out next Monday to survey some parts of Westphalia. He has not yet begun to print any part of Von Buch's works, but will begin before winter.

Humboldt is busy with the fourth volume of " Cosmos," and Von Decken told me that the chief subject of it will be volcanic action.

Pertz took me to the Academy last Thursday, when Heinrich Rose read an analysis of a rock salt from some new locality. Encke was in the chair, and I was introduced to him, and some others of those great men.

Dove, Mitscherlich, Müller the physiologist, the two Rose's, Bœck, Bopp, Lepsius, Jacob Grimm, Trendelenburg, Braun the botanist, were among the number, and I was pleased to see the quiet, unpretending way in which the business was conducted, in favourable contrast with the velvet cushions and Mace of our Royal Society. Lepsius shewed us yesterday the Egyptian Museum, and through the new museum of Grecian Art, where we found Kaulbach at work on the adornment of the walls.

The heat here is great, in the middle of the day from 88° to 90° in the shade, but the weather is beautiful, particularly in the evening.

Of politics I hear next to nothing; except what is in the newspapers, I learn nothing. The little I have heard is decidedly anti-Russian, but there does not seem any desire for anything beyond a strict neutrality. I have never been able to see any sufficient ground for Prussia doing more.

My best love to my dearest Mary, I need say nothing of my companions for they will tell their own story. Nora is evidently happy, and certainly has a devoted husband.

Ever, my dear Charles, affectionately yours,

LEONARD HORNER.

_____

To his Daughter.

Berlin, *July 30th,* 1854.

MY DEAREST FRANCES,—I rejoice to hear of your weather being so beautiful as to give you full enjoyment of your lawn and garden. We have been particularly fortunate in that respect ever since we left home. I thought before I left London that the expectation that Prussia should have declared war against Russia was very unreasonable. The danger to this country of Russia extending her power in the south of Europe, is very far inferior to what it is to

Austria, France and England, and nothing short of imminent danger to the country would justify so enormous a price as she must pay by going to war. Look at the immense frontier she has, conterminous with Russia, not less than eight hundred English miles, and think what a force must be required to defend such a frontier, assailable at almost every point, and which she must herself defend. Except the officers, Prussia has no *standing* army in the sense of the term as applied to France and England, for none of the common soldiers are such for more than a limited time. Her *nominal* standing army is two hundred thousand, but in time of peace only two-thirds of that number are under arms, and they are composed of young men from twenty to twenty-three years of age, who serve for one or three years, the former fitting out themselves, the others not. When war is declared, the first step is to call out the remaining third of the two hundred thousand, that is the reserve, and then comes the Landwehr, which amounts to four hundred and fifty thousand men. The cost of putting this force of six hundred thousand men in marching order, is estimated at eighteen million of thalers, and then comes the subsistence and the expense of moving this immense army. But the money directly paid by the country is nothing compared to the loss by the immediate cessation from productive labours, of four hundred and fifty thousand men, and of the change of habits from their previous occupations, when at the return of peace they are discharged. You will thus see what a momentous question the Government of Prussia has had to decide, and can we be surprised if they seek by every means consistent with honour to avert so terrible a calamity?

I had the great honour yesterday of a visit from Humboldt. He is eighty-five, and in full possession of his faculties, and a most agreeable talker. He sat three-quarters of an hour with us.

The Museum here is a collection of treasures of many kinds,
and we saw the Egyptian part first under the guidance of
Lepsius, but I have been twice there since, for it is very
instructive. We had a very pleasant excursion last evening to
the neighbourhood of Spandau, where there is some very
pretty lake scenery, near the village of Pickelsdorf. I am
particularly struck with the good taste and beautiful execution
of the external architecture of a large proportion of the new
private dwellings and the public offices.   Unter den
Linden is a most beautiful street, and the statue of Frederick
the Great at one end of it is truly grand. I must not omit
to mention the noble public library of six hundred thousand
volumes, open on the most liberal terms ; the books, with the
exception of a comparatively small number, being lent out.
The arrangement, entirely the work of George, is admirable.
As a proof I may mention that I asked to see a book a day or
two ago from one of the assistants ; it was one by no means
in common demand, and in five minutes it was in my hand
traced out by the admirable catalogue. Sir Henry Bunbury's
last publication has been for ten days in the catalogue, so
that you see they are well up in new books.

We propose staying here till the 17th when we go to
Dresden, and we shall be in Paris on the 27th, and intend to
remain there ten days, and to be home on the 8th September,
but I shall only be a week at home, for I must go without
delay to my work in the North.

Our affectionate love to Charles.

God bless you, my dearest Fan,

Your affectionate father,

LEONARD HORNER.

To Charles Bunbury, Esq.

Berlin, *Sunday, August 6th,* 1854.

MY DEAR CHARLES,—Last evening I was at a meeting of
the Geographical Society to which I was invited by the

President, my old friend the celebrated Carl Ritter, and also by
my friend Mitscherlich, the Professor of Chemistry, who called
for George and me. As we had had prior engagements, we
did not get there until near the end of the *meéting for business*,
when Dove was giving an account of a new work on glaciers,
illustrated by beautiful drawings. But we were not too late
for the social meeting, for while we dine before our meeting
at the Geological Society, they sup after theirs. I sat next to
Ritter, with Ehrenberg on my other hand, and I had much
pleasant talk with the geographer. He told me much about
Kieppert, whose lately published maps of the Caucasus and
other Eastern countries you may have seen. Kieppert is an
accomplished scholar, who has travelled much in Eastern
countries. He passed a long time in Asia Minor, where
with four other Prussian officers, he made many accurate
surveys, and with those he has constructed a new and very
detailed map of that country, He has been elected a member
of the Academy. On Saturday we went to the Atelier of Rauch,
where we saw many beautiful things, among them a bust of
Humboldt, on which he is now at work, that is, on the marble,
for the model is finished. But we were most interested in
seeing the sculptor himself, who has just returned from Italy,
and a very fine venerable head he has. We are all much
pleased with his great work here, the bronze statue of Frederick
the Great in the street *Unter den Linden*. The figure of
Frederick and the horse are admirable, although so colossal
that the head of Frederick is forty feet from the ground. The
pedestal, which is covered with figures of the great men of his
time, four of whom are on horses as large as life, appears to
me much overloaded. We saw many productions of his pupils
that shewed great artistic skill, both in the conception and
execution.

We passed two hours with Ehrenberg some days ago, seeing
many of the wonders he has made known by his microscope,

and this morning I was with him three hours. You may
remember my telling you that I asked Mantell to examine
several specimens of Nile mud with his microscope, and that
he told me he had not been able to find a single organic form.
On the recommendation of Ritter, I sent portions of the same
specimens that Mantell had examined, to Ehrenberg, and he
sent me the result, which was sixty-seven different forms, partly
animal but chiefly vegetable bodies. This startled me a good
deal, as it did all our naturalists to whom I told what Ehrenberg had discovered, where Mantell could see nothing. I was
therefore anxious to see with my own eyes what Ehrenberg
had described, and on expressing my wish, he most readily
complied. He first shewed me the engraved figures in a work
that will appear in the course of a few weeks, and then the
substance, or rather the organism I should say, itself, and I
saw about a dozen of them taken at random, so clearly, that
there is no reason to doubt his being right in all the rest.
Indeed no one could see and talk to the man, and examine
his work in detail, and not be convinced of the perfect honesty
of the observer. There is an openness and simplicity of
manner, and a calmness that inspire great confidence in him.
He shewed me the organic bodies, which he found in mud
brought up from the sea bottom of the Atlantic between
America and the Azores, at a depth of 12,000 feet, most
distinct and peculiar forms, and these not fossilized, but
bodies with shells, which when removed by acid, left a flaccid
skin. Now it was thought until these discoveries, that there
could be no living creature below a depth very far inferior to
that, I think the maximum was under 1,500 feet; Elie de
Beaumont stated it at much less, and if I mistake not, Edward
Forbes in his Memoir on the Ægean Sea, makes the maximum
less than 1,500. These then are very important results,
which Ehrenberg has arrived at, and he has recently made a
still more important discovery, for he has found organic forms,

polythalamia in a green sandstone, that underlies the lowest
of the Silurian beds, from a locality near Petersburg. See, as
to the green sandstone, Murchison's new Silurian volume,
page 373. It has always appeared to me exceedingly improbable that the land which supplied the materials of our
sedimentary deposits, was without animals and plants, and
that the sea which received the materials was untenanted.

These last discoveries of Ehrenberg go a great way to
support that view, and it is not at all improbable, that even
in the metamorphic strata, organic forms may be discovered,
when the action was not sufficient to obliterate the siliceous
coverings of the microscopic animals, as is the case with
sedimentary deposits that have been thrown out in fragments
from active volcanoes.

We went a party of eight to Potsdam. George could not
go, as he was engaged to dine with Encke, who gave a great
dinner on the occasion of his laying down the annual office of
Rector Magnificus of the *Senatus Academicus*, or rather of the
University, in which high office he is to be succeeded by
Mitscherlich. We did not go to the Palace in Potsdam, but
confined ourselves to *Sans Souci*, the palace and gardens.
The fountains were playing, and there is one jet that reached
a height of 150 feet, the finest water-work I ever saw. The
sun shone full upon it, and we had a fine rainbow. The
gardens are extensive and beautiful, with a profusion of
marble statues, vases and benches, and all in the most perfect
order. There were many companies of people, many of them
the humblest classes, and all observed the greatest propriety.
Not a single policeman or soldier on duty was to be seen in
the whole place. We saw the interior of the palace and with
interest, the rooms and many memorials of the great
Frederick.

August 8th. It has somewhat surprised me to hear so little
said here on the great question of the war, but with few ex-

ceptions the persons I have seen are mostly engaged in scientific pursuits, and I have rather talked with them on their own subjects. I have not yet met with one person who is not decidedly anti-Russian, so that should war be decided upon, it will be a popular movement in this country, so far, at least, as feeling against Russia is concerned, however heavy the sacrifices must be felt. But from occasional conversations I have had with men of a liberal but at the same time moderate turn of mind, I see that there is much dissatisfaction with the present Government, especially on account of the increased influence of the bigoted, narrow minded clergy, of which party Hengstenberg is the leader, and he has been placed in a situation where he can do the greatest mischief, for he is in the Commission for the examination of candidates for the office of teacher in the gymnasia, with the sole power of admission or rejection on his religious grounds. He is supported by Raumer, the Minister of Public Instruction.

Believe me, my dear Charles,

Affectionately yours,

LEONARD HORNER.

————

To his Daughter, Madame Pertz.

Lancaster, *September* 18th, 1854.

MY DEAREST LEONORA,—It is now one calendar month and a day since I last saw your dear face on the platform at Berlin, when we parted after nearly five happy weeks together. I daily think of the pleasure our visit to Berlin gave us, and cherish every minute recollection, that I may preserve a familiarity with all that is going on at No. 40, Behrenstrasse, and among the friends you see. I hope you will continue to give detailed accounts of all that goes on at home that we may be

as much together as possible. It has been a very great
pleasure to all of us to have dear Hermann in our house.

I left home last Thursday morning, and reached Carlisle
soon after five, I had for my companion for about forty
miles, William Greg, the author of many excellent articles in
the *Edinburgh, Quarterly* and *Westminster Reviews,* which
essays have lately been collected in two volumes. He is the
brother of Samuel Greg, and an old friend of mine. He was
a cotton spinner, but never with his heart, so he left off the
uncongenial occupation, and now leads the life of a *literateur,*
residing on the banks of Windermere. I was all Friday at
Carlisle, I found the Taits absent in Scotland. Yesterday
was not at all *langweilig* for I had an interesting book, Dr.
Henry's "Life of Dalton," a copy of which he sent me.
Dalton was a man of whom Sedgwick said, that from the
time he came from his mother's womb the God of Nature had
laid his hand upon his head, and had ordained him for the
ministration of a high philosophy. To-day I have been
visiting mills in the town and neighbourhood; I shall be
occupied in the same way to-morrow, and hope to arrive at
Liverpool in the evening, to attend the Council of the British
Association, which meets early next morning. I daresay you
have heard that the Henry Romillys have given Charles and
Mary and me their house during the Association, which will
be very agreeable for me.

Your affectionate father,

LEONARD HORNER.

To his Wife.

Liverpool, 26*th September,* 1854.

Yesterday Charles occupied the Chair in the Geological
Section, as Forbes had to read a paper in the Natural History
Section. Murchison gave us a lucid and interesting account

of his recent observations in the Hartz and the Thüringer Wald. I went to hear Colonel Sabine's paper on terrestrial magnetism, which was to be delivered in the great hall. I called on Dr. Fowler on Sunday, and as he wished to hear the lecture, I undertook to find a fit place for the good old man ; so I found him and Mrs Fowler in the Hall, and leaving her where she was, in a very good place, I took him to the platform, where he was kindly welcomed by Colonel Sabine and placed in the best position. He enjoyed it much, and although eighty-nine, is as alive in mind as ever. 1 spoke to Mrs. Jameson, and Pillans, and Lord Harrowby, who was still suffering from hoarseness, that Murchison had to be his mouthpiece.

I am glad to hear that Lady Bell enjoyed her visit to Mildenhall and Barton. I am sure she would give as much pleasure as she received.

To his Daughter.

Manchester, *October 8th,* 1854.

MY DEAREST MARY,—God be praised that I can again wish you and myself joy at the return of your natal day, which will see you to-morrow, I fervently trust, in perfect health, to continue a blessing to your dear husband and all of us ; for a blessing you have been to all connected with you since the day of your birth. I dwell with pleasure on the remembrance of the happy week I passed with you and Charles at Liverpool, to me the best part of the meeting, for it was long since I had been so long under the same roof with you by night and by day.

What a state of anxious suspense must those persons be in who have sons and brothers and husbands in the Crimea, until the fearful list of casualties is made known ! I fear too that long after the battle of Alma there will be much hard fighting and much bloodshed, especially if it be true that the Russian General Ochsen Sachson is coming from Odessa with a great force. I hope we shall soon hear of more troops being sent from France, for we have no more to send. If Austria allows the Russians to send forces from Odessa into the Crimea, it is tantamount to siding with Russia, and I hope that neither France nor England will submit to it, and that France will send one hundred thousand men into Lombardy, and that we shall take possession of Trieste and Venice.

Joanna is very well, and you know how happy she always makes your Mamma and me when she is with us.

God bless you, dearest, dearest Mary. Give my love to Charles.

Your affectionate father,
LEONARD HORNER.

To his Daughter.

Manchester, October 14th, 1854.

MY DEAREST FRANCES,—I was very happy to hear that your minds had been relieved about Major Bunbury. When we heard how dreadfully his regiment had been cut up, I could not help rejoicing that he had been prevented from landing in the Crimea. There are no doubt in his mind many regrets that he was not in the great battle which has won such laurels for our brave troops. But it is to be expected that other occasions will present themselves in a few months, after he has rejoined his regiment, when the determined spirit and skill of our officers and men will be productive of other great and glorious results, for the fall of Sebastopol will not bring the mad Czar to his senses, and much more blood will be shed by the wicked obstinacy of that scourge of Europe.

I am going on in my usual routine, and my health is quite equal to my work. I will desire our Clerk at the Factory

Office to send you a pamphlet on the late strikes, so sensible,
so good in all respects, that I am doing what I can to give it
extensive circulation.   I have got a bookseller here to get a
supply, and drew up a handbill for him, which he had printed
for distribution among the mill owners.   I enclose a copy of it.
The author of the pamphlet is Mr. Samuel Robinson, brother
of the late Lady Heywood; the author of the notice in the
Economist is William Greg.

With our united love to Charles, Harry, your sisters, and
kisses to the darling boys.

I am, my dearest Frances,

Your affectionate father,

LEONARD HORNER.

To his Daughter.

Manchester, 29*th October*, 1854.

MY DEAREST SUSAN,—I last night finished reading your
Memoir of Kossuth.   That I had not done so long ago, you
will not, I am sure, impute to any want of interest in an enter-
prise of yours ; when it came out we were on the point of
leaving for Berlin, and being desirous of taking advantage of
the German atmosphere and of knowing something more
than I did of Frederick the Great, what time I have for
reading while I was abroad, was exclusively given to German
books.   You know how much I had to do while I was at home
and how little time I had for reading during the time of my
official duties.   Three-fourths of the books I brought in the
expectation of reading them, have not been looked at, and we
hope to be at home next Tuesday.

I have been *greatly* interested by the Memoir, and by
becoming better acquainted with him, my admiration of
Kossuth is *greatly* increased.   Your part is admirable ; the
narrative throughout is clear and readily carries one on ; the

style appropriate, and where you have indulged in any reflections, in any sentiments of your own, they are expressed with
courage, moderation and dignity.

It is impossible for any candid person to rise from the
perusal of this work without admiration of the purity of
motive, and the lofty enlightened patriotism of Kossuth, his
perseverance, his indomitable spirit and courage and patience
under the most severe trials, and without lamenting that so
just and noble a cause was shipwrecked by the jealousies,
irresolution and feebleness of many of those with whom he
was forced to act ; to say nothing of the perfidy of the great
traitor Georgy.   When I take into account the enormous
difficulties with which he had to contend, and that he never
had any instruction or experience in the art of war, I cannot
charge him with a single fault from beginning to end of the
struggle ; even his leniency to Georgy, considering all the
circumstances of the case.   But what shall I say of the
Government of Austria ?   I cannot find words to express my
horror of their conduct, or of the monsters who were the
instruments of their perfidy, and savage cruelties.   All the
horrors of the Hungarian war are traceable to one man—
Metternich ; his endeavour, systematically pursued, to stifle
the freedom of Hungary, and bring it under the yoke of
domestic rule, just as I trace the loss of freedom to France to
Guizot, resisting the moderate freedom of speech on political
subjects indulged in at the banquets, and refusing to allow
the very moderate measure of reform in the Chamber, to be
brought about by increasing the number of electors and
expelling some thirty or forty of the employés from the
Chamber.   Louis Napoleon should erect a statue of gold to
Guizot for his elevation.

I cannot help having a strong feeling of regret that by
publishing your work anonymously, you should be deprived of
the credit and honour which so many would be willing to

admit to be due to you; but perhaps it is right, although your Mamma's arguments have failed to convince me.

Much as we delight to have your sweet smiling face near us, do not deprive yourself, nor Charles and Mary, of the pleasure of your intended visit to Harley Street.

God bless you, darling. Love to all.

Your affectionate father,

LEONARD HORNER.

To his Daughter.

17, Queen's Road West, *November 25th*, 1854.

MY DEAREST LEONORA,—I am sure you would warmly sympathize with us in the great loss we have sustained by the lamented death of Edward Forbes. During the week at Liverpool, when I was with him daily, he was remarkably well, and was looking forward with much interest to his winter's work at the University. He appears to have caught cold afterwards, during a dredging expedition with Sir W Jardine, and the fever fell upon a weak part, the kidneys, and he suffered excruciating pain. His loss is irreparable in his department of science, and in his social circle he will be long missed. For Edinburgh University the loss is most serious, for there is no man to be had the least to compare to him in qualifications for the Chair of Natural History, for he was eminent as a zoologist, a botanist, and a geologist. At the meeting of the Philosophical Club the day before yesterday, his loss was greatly felt and deplored. I was in the Chair, and Professor Bowman of King's College, an eminent surgeon and physiologist, when I was talking to him of my very agreeable visit to Berlin, spoke to me in very high terms of Professor Müller, and we rejoiced together at the distinction conferred upon him by the Council of the Royal Society, in awarding him the Copley medal, the highest distinction they have to bestow. When you see him and Mrs. Müller remember me very kindly to them, as also to their accomplished daughter. When is their son coming here?

I have nothing very particular to tell you of my own occupations since we returned from Manchester. What leisure I can command from my official duties has been in great part occupied by the resumption of the arrangement of the multifarious piece by piece information sent me by Hekekyan Bey. I have recently had a letter from him giving me a brief account of last summer's operations, during which seventy-two pits have been sunk in the alluvial land, in the parallel of Heliopolis, between the Arabian and Libyan ranges of hills. When I get the details of these last operations, I shall be able to make up my account of all that has been done. You may tell this to Baron von Humboldt and Professor Lepsius, with my best regards. . . . .

Ever, my dearest Leonora,

Your affectionate father,

LEONARD HORNER.

CHAPTER XI.

1855—1856.

To his Daughter.

Manchester, *9th April*, 1855.

MY DEAREST NORA,—I have read little more this time than
one book, and that has been *Stein's Leben*.* I am now in the
middle of the third volume and it is with difficulty and regret
that I lay it down. I have as yet found everything admirable
in his whole life, and if he had not been of the warm
temperament he was, he never could have struggled against
the dispiriting things he was constantly meeting with. He
was a great enlightened statesman in the noblest sense of the
word, with the utmost and purest devotion to his native land,
and by that I mean, *das ganze Deutschland, nicht allein Preussen.*
Then his letters to Frau von Berg, to the Princesses Louise and
Wilhelm, and above all, those to his wife, shew that he was a
man of the warmest affections, and had he given way to his
natural feelings and rich resources, he would have spent many
of those years which he gave up to his country, in peaceful,
domestic life. Had he been placed in the same circumstances,
he would have acted as Washington did, and greater
praise I cannot bestow upon him. The interest of the
narrative is admirably sustained by his biographer, whose
own reflections are no small part of the value of the work.
I am glad to hear that he looks forward to being soon able to
set about the abridged edition.

I yesterday received a large packet from Hekekyan Bey ;
further report of his operations, which will enable me to make

* By Dr. G. H. Pertz.

some progress in my memoir as soon as I return home. The Council of the Royal Society have; ordered the paper I presented in February to be printed in the *Philosophical Transactions*, which probably ensures the whole of my memoir appearing through the same channel. What I have given will constitute the first part of the memoir, and I contemplate two more. Besides what I have written for the Royal Society, in which I have confined myself to my geological objects, the excavations there brought to the surface reveal many curious buried remains of human art, of which I shall give an account through some other channel.

I am, my dearest Leonora,

Your affectionate father,

LEONARD HORNER.

To his Daughter.

Manchester, *April 22nd,* 1855.

MY DEAREST MARY,—It has made me very happy to learn that you and Charles passed your time so agreeably at Berlin and Paris.

During all the time I have held my office, I have never passed so disagreeable a time as during the present sojourn in Manchester, from the injustice I have met with from many of the leading mill owners, who have been roused into violence and passion, by having been called upon to observe the law about fencing dangerous machinery, which would entail some trouble and expense, and they have vented their rage upon me.

I have been much gratified by hearing from Murchison that he has agreed to be the successor of poor De la Beche. He is unquestionably the best man for it, and will be, I doubt not, eminently useful. He is so kindly disposed that he will be sure to please the working men in the School of Mines, and will be very active in forwarding the scientific objects and prevent their being obstructed by the preponderance of the department. Murchison, in his letters to me, speaks in the warmest terms of gratitude for the way in which his friends have shewn their regard for him, and the testimony they have given of their sense of his merits.

I cannot tell you with what disgust I have seen the fulsome adulation that has been poured forth on Louis Napoleon. It is a national disgrace. If he had been a Washington, ennobled by every exalted virtue, political and moral, more could not have been done.

I am sure that every right thinking man in France, in place of thinking more highly of England, will despise the praise and worship bestowed on a man they know to be so thoroughly undeserving of it. But I do not the less admit that, within due bounds, any civility to the person the French people have chosen for their chief Magistrate, calculated to strengthen a good understanding between the two countries, may have been justifiable, and even more, but it is the uncalled for worship of *those* unconnected with Government, and of the *newspapers*, that disgusts me.

Your affectionate father.
LEONARD HORNER.

————

To his Daughter.

London, *May 11th*, 1855.

MY DEAREST NORA,—I received yesterday the notice of the new volume of the *Monumenta*. It must be gratifying to George to see his great labour so justly appreciated; and many testimonies to the same effect I have no doubt he has again and again received, from all parts of Germany, as well as from other countries, for the work illustrates the history of most of the countries of Europe. I do not find so much time for *Stein's Leben* as I did in the evenings at Manchester, and I am only advanced but a short way in the fourth volume. It is terrible how one's time is cut up into shreds and patches in busy London by the incessant calls of all sorts. I am reading, just now, a life of a very different sort, that of Dr. Thomas Young, by the Dean of Ely. I read it by snatches, as I can command a half hour. If it is not in the library, George would find it a valuable book to possess, together with the edition of Young's works, now in course of publication, Murray is the publisher. The life is an admirable lesson of what unflinching perseverance in study can accomplish.

I have met twice, at the Philosophical Club, a very agreeable Berlin Professor, Emil du Bois-Raymond. He is lecturing at the Royal Institution, and he is very highly thought of by our most eminent physiologists. He is to be with us next Tuesday evening.

I have just had a handsome present made to me by Admiral Smyth and Dr. Lee of Hartwell House, in Buckinghamshire, of the Admiral's description of Hartwell House, a fine old mansion, which the present possessor has enriched with many valuable collections of antiquities and works of art, and to which he has added an observatory with very fine instruments. Ask Mr. Encke if he has got a copy of "Ædes Hartwellianæ"? Besides an account of the observatory and of the observations made there, there is a representation of his own comet as seen there. The work is not published, but printed for private distribution. If there is not a copy in the library, I will do my best to get one.

If you have an opportunity, present my most kind and respectful regards to Baron von Humboldt. He was a subject of much conversation at the Philosophical Club yesterday, and Colonel Sabine told us of an admirable likeness of him sitting in his room, by Hildebrand, lately come out.

Your affectionate father,

LEONARD HORNER.

To his Wife.

43, Inverleith Row, Edinburgh, *July 22nd*, 1855.

Here I am safe and sound and in most comfortable quarters. I was in good time for the train, but not too soon. When the Bass and North Berwick Law and old Arthur's Seat came successively in sight, I felt as I suppose the Swiss feel when they return to the sight of their mountains. Emerging from the station up to Princes Street, I saw the new National Gallery on the Mound—perfectly beautiful. What a group is presented by that first sight of the town to a traveller from the South. I found Pillans perfectly well, looking younger than when I last saw him. I met a most hearty welcome, had a good tea, and a long talk until about eleven o'clock.

I found a kind note from Murray, who is coming to breakfast here, as do Tom Jackson and his wife.

Afternoon. After finishing letters to you and others, and reading some pages of Herder's life, it was nine o'clock, and first arrived Henry Rogers of Boston, the geologist, whom I was surprised to find in this country, having left Boston a month ago ; then Dr. Schmidt, then Lord Murray, and Mr. and Mrs. Jackson. . . . . . A pleasant talk on all sorts of subjects went on for two hours which was prolonged for two hours more, in the library upstairs. Murray, Pillans and I, then went to the beautiful Botanic Garden, which is in full luxuriance of leaf and flower. Murray soon went, but Pillans and I prolonged our walk, accompanied by Mr. Macnab, the very intelligent head gardener. We mounted on the top of one of the conservatories to enjoy the very fine view of the town, which we saw to advantage as the day was fine. Government have granted £6,000 to build a new Palm House, and it is to be begun forthwith. Pillans and I then took a cab and drove to Coates House. Jane met us at the door and was a good deal overcome.* She went to her mother, who said she would see me for a few minutes. Poor dear Mrs. Cockburn was sitting on the sofa in her widow's cap. She was much affected. I stayed a very short time, and she made me promise to come back.

Monday. From Coates House I went to call on Playfair, but did not see him, as he goes to bed during a part of the day. He is become in body, that is in his limbs, almost helpless, but his mind is entire; he takes on no new business, and is winding up with the National Gallery on the Mound, one of his most perfect works. I drove across the Mound to get a nearer view of the new building, and thence to Adam's Square to look at the statue of Watt. It is very good, but the pedestal sadly defaced from the absence of a railing. There are three steps on which the dirty children of the Cowgate sit, leaving their marks behind them, and out-door preachers make use of the steps as a Rostrum. I shall see what I can do with the Town Council to get a railing put round.

I called on the Murrays, saw them for a quarter of an hour, and met Lord Minto, who gave me a hearty invitation to Minto. Then I went to Brodie's studio. He has done a most beautiful bust of Cockburn for James Craig, as large as life, unquestionably the most striking likeness I have ever seen of him, indeed it is perfect. Then I called on Mrs. Forbes, whom I found at home, and upon the whole cheerful, a very pretty house. Then to Dr. Murray to talk about the School of Arts, and then to the Royal Society's Rooms, in which I am now writing. I had not been long here, when John Russell came in, who most heartily greeted me, and we had a long talk, then Dr. Hannah, our late Rector, and then Professor Swinton, so we had much talk about the Academy. I am now going home to dine *tête-à-tête* with Pillans.

*Her father, Lord Cockburn, having died since she had last seen Mr. Horner

To his Wife.

Edinburgh, 24*th July*, 1855.

I was at the Academy from ten till three, and have been much gratified with everything. It never was in a more efficient state. I have met with the most hearty welcome from all my friends connected with it. To-morrow I breakfast with the Directors and the Club, and the head boys of each class ; at twelve there is the Exhibition, and John Russell and I have agreed to go in arm-in-arm, the two left of the original Directors. Then we have the dinner afterwards. I have been sitting with Sir William and Lady Hamilton, who are both well, he looking a great deal better than when I saw him last.

July 25*th.* This morning I breakfasted at Barry's Hotel with the Academy Club, the Directors and the dukes of the seven classes, very fine boys all of them. I sat between Mr. Lonsdale, the Bishop of Lichfield's son, who came down on purpose as an examiner, and Professor Campbell Swinton.

At eleven I went to the Academy, where I enjoyed the very pretty sight of so many fine gentlemanlike boys in their holiday clothes, walking about the playground, and before going into the Hall I had much talk with many old Academy friends. It was a very fine day. The room was crowded to excess, you know what a pretty sight it is. The prizes were delivered by the Solicitor-General, Maitland, who made a most excellent speech. But I must stop, for it is time to go to the dinner. God bless you all.

———

From Charles J. F. Bunbury, Esq.

Mildenhall, *July* 29*th*, 1855.

My Dear Mr. Horner,—I thank you very much for your memoir on Egypt ; I have read it through with attention, and think it very clear, interesting, and instructive. I am par-

1855.] HOOKER'S FLORA INDICA.

ticularly pleased with your preliminary view of the physical
and geological structure of Egypt in general, which appears
to me a very masterly piece of physical geography. Your
researches have been so well planned, and carried out on so
noble a scale, with so much perseverance, and such elaborate
care, that I am extremely glad they are to be given to the
scientific world through a medium which will attract attention
to them, and do them justice. I shall be eager to see the
subsequent memoirs in which you will develope the theoretical
conclusions that you deduce from the facts hitherto ascertained.

I am much obliged to you also for letting me see the
examination papers of the Edinburgh Academy. I congratulate you on having found that Institution in such a
flourishing state; it must be a great satisfaction to you to see
its continued success and usefulness.

I am exceedingly pleased with Hooker's "Flora Indica,"
which I am studying carefully; it is a most masterly work,
and will, I think, not only support but even heighten his
previously great reputation. I am also going carefully through
Geinitz's new work on the fossils of the coal formation in
Saxony, and I find it so important for the knowledge of fossil
plants, that I intend to send a somewhat full account of it to
the Geological Society's journal. His views appear to me
remarkably sound, and I can answer for the correctness of his
plates and descriptions, having seen the original specimens in
the museum at Dresden.

Ever your affectionate son-in-law,
C. J. F. BUNBURY.

To his Wife.

Drumkilbo, July 31st, 1855.

I wrote to you yesterday after my arrival here, and this
morning I received yours and Charles Bunbury's kind letters.

I am much gratified by his favourable opinion of my Nile
paper, because I am sure he is far too sincere to say more
than he really thinks.   This is really a very pretty place; it
is surrounded by very fine trees and pasture land, and near
the house a great abundance of evergreens, many of very large
size.   Nothing can be more kind than all the Miss Lyells
and Tom are.   It is quite nice to see Tom with his nephews,
he is so kind to them, and he seems to have quite won their
hearts.   I was looking out of the window just now to see
Leonard and Frankie in their turns riding on Donald, a most
beautiful Shetland pony, with a tail almost touching the
ground, and nice new saddle and bridle adapted to the size of
the children.   The dear pets are evidently in a state of great
enjoyment.   Arthur is improving in intelligence, smiles most
sweetly, and he now walks with considerable ease.

August 2nd.   Yesterday, soon after breakfast, Sophy,
Katharine, Leonard and I, with Tom, set off for Shiel Hill.
After a great deal of rain, it was a fine day.   The distance is
sixteen miles, we were there three hours, and got back soon
after seven.   It is a very pretty place indeed, the situation on
a sort of table-land, with masses of wood in the distance,
overhanging the river Esk.   The ground and gardens near
the house very nicely laid out.   Tom has made a considerable
addition to the house, and a most comfortable one it is, and
excellently well furnished.   Leonard was in high spirits and
most happy and good all the day.   As we passed through
Kirriemuir we paid a visit to Bell Lyon, now eighty-seven,
she was delighted with Leonard, saying he was the fourth
generation she had seen, she went out and bought cakes for
him.

Give my love to all and believe me,

Your affectionate husband,

LEONARD HORNER.

To his Wife.

Ambleside, *August 5th*, 1855.

I wrote to you in the middle of the day on Friday, after which I went a round of country mills and came home soon after five o'clock and spent a quiet evening alone. My reading was Herder's Life, a very interesting piece of biography. I am little acquainted with his writings, but the account of his life makes me desirous of reading some of them. He was a very remarkable man, who from a very humble origin, the son of a small village schoolmaster in East Prussia, not far from Königsberg, raised himself to great literary distinction and influence. You will remember our seeing his house and statue at Weimar last year. Yesterday morning I visited some mills at Kendal, and at two o'clock, by rail and coach, arrived here. After my arrival I called at the Davy's, and saw him for a few minutes, as they were preparing to go out in the carriage, came home and dressed and got to Mrs. Fletcher's at Lancrigg by five o'clock. The Davys dine every Saturday with Mrs. Fletcher and she every Wednesday with them. She had a dangerous attack of bronchitis six weeks ago, when she was nearly dying, but she is now very well, and although three years have told upon her, and she appears more feeble in body, in mind she is as fresh as ever, taking a warm interest in all public events, and a steady, good Whig, disturbed by what she considers the unmerited attacks upon Lord John Russell, in whose honesty she has unshaken reliance, and in this we were quite agreed. Her admiration of Kossuth as great as ever, and when I told her that I would let him know that he has so devoted a worshipper in the north, she said it would be an honour to her to be so termed. She desired to convey her respects to Mr. and Mrs. Pulzsky. There was a family party, Dr. and Mrs. Davy, their son and two daughters, and Mrs. and Miss Taylor. To-day, at four o'clock, I dine with the Davys. To-morrow I leave early and expect to reach Preston between twelve and one o'clock.   There has been much rain, yesterday frequent showers but much bright sunshine, and I had the latter most brilliantly in going to Lancrigg, and you know how exquisite the scenery is by Rydal and Grasmere Lakes, and as seen from the terrace on which Mrs. Fletcher's house stands.   I am going to call on Miss Martineau to-day, she is in the same precarious state of health, but goes on writing her auto-biography.   The Davys and Mrs. Fletcher have had no intercourse with her since the publication of her letters ; you will remember the capital letter Mrs. Fletcher wrote to her on that occasion.   Her brother, in a review of the book in the *Westminster Review*, lamented her grievous folly, which has broken off her intercourse with him.   He was here a few days ago and called upon her, and on holding out his hand to her she drew back.   Mrs. Gaskell was here a few days ago visiting her, she is collecting materials for a memoir of Miss Bronte (I think that is the name), who published under the name of Currer Bell.

I am glad you had the pleasure of a visit from Sedgwick, he is very entertaining, and an excellent man.

To his Daughter.

Manchester, *August 12th*, 1855.

My Dearest Leonora,—I congratulate you and George on the distinguished mark of his Sovereign's admiration of his great historical work now brought to a close.   Would that he would read and ponder over the wisdom these volumes contain, and call to his aid Ministers such as Stein, if such are to be found.   Another like him does not probably exist : heaven does not send upon earth such spirits except at distant intervals, but there may be men who hold the same enlightened views of civil and religious liberty, whose sole thoughts are directed to the happiness and prosperity of the governed, as those of Stein's so readily were.

Your affectionate father,

LEONARD HORNER.

To his Daughter.

Manchester, *Sunday Night, September 2nd,* 1855.

MY DEAREST MARY,—The visits you paid in Edinburgh were some of them very painful to you, but I am sure they gave pleasure where you felt most pain, by the sad changes connected with them, for the sun shone out for a time while you were with your widowed friend. The visit to the Dean Cemetery is to us indeed a very sad one, so many lying there so lately gone, that we have them strongly before us in their living forms. The monument to Jeffrey is good, so is that to Forbes; I trust that good taste will be shewn in the monument which is to mark the spot where the remains of our dear loved Cockburn rest.

Your kind-hearted visits to the poor Milnes of Gorgie, to Mrs. Gillies, to Miss Maclaurin and Miss Franklin, are only expressive of that combination of benevolence and duty which have been characteristic of you through life, and which have been the gems in your nature which your dear Mamma and I have fondly dwelt upon as the brightest jewels. You do not appear to have seen Pillans, which I am sorry for, as I am sure it would have given him pleasure.

I have offered to dine with Kay Shuttleworth at Gawthorpe Hall on Friday. We shall leave this place I expect on the 12th, and as your Mamma was so anxious to pay the Dean of Hereford a visit, we shall go there on Wednesday, by Chester and Shrewsbury, and get to Hereford between seven and eight in the evening.

I have just finished Hugh Miller's Schools and School

masters. I have read every word of it with interest and sorry I was when I got to the end of the volume. Perhaps "every word" is saying too much, for I could have dispensed with the poetry. It is a very remarkable book, and curious and valuable as a picture of humble life in Scotland.

I cannot understand the reason why he so shuts himself up at Edinburgh, and avoids the society of many who would have great pleasure in his friendship, and would assuredly add to his happiness if once he got over his retiring habit. I shall be glad to hear some account of your and Charles's late interviews with him.

Your affectionate father,
LEONARD HORNER.

To his Daughter.

Manchester, *April 20th*, 1856.

MY DEAREST SUSAN,—I have been greatly interested by hearing read your letter of yesterday, with so masterly a sketch of Kossuth's lecture on the *Concordat*, which I should have listened to with no less interest than you did, but should have been deplorably beaten had I afterwards entered into competition with you to give an account of it. It has occurred to me that Kossuth would render good service to the cause of religious freedom if he were to publish these lectures. I would not have them published in the form of a pamphlet, for that kind of publication rarely pays the cost, but in the form of a cheap duodecimo book. So strong and so widely spread is the Anti-Catholic feeling in the country, and particularly so at this moment, that there could hardly fail to be an extensive demand for it. Now although the Anti-Catholic feeling is mainly an ignorant and discreditable one, flowing from a narrow intolerant spirit, it is to a great extent against the priesthood and priestcraft, and it is therefore legitimate to strengthen *that* intolerance by all means in our power. It is not because of the belief that I think the Roman Catholic the worst form of Christianity, because all or nearly all forms have some irrational dogmas, the Roman Catholic more than any other; my abhorrence of it arises from the audacious dominion the priests exercise over the minds and conduct of their subjects. I think that such a publication would to a considerable extent increase the respect so justly due to Kossuth, and change the opinion of many who so ignorantly judge of him.

My dearest Susan,

Your affectionate father,

LEONARD HORNER.

To Dr. and Mrs. Pertz.

17, Queen's Road West, *June 12th*, 1856.

MY DEAREST GEORGE AND LEONORA,—I cannot be content to send by another, although that other be my other and my better half, my heartfelt thanks for your affectionate remembrance of the 50th return of that blessed day when I became richer than all the gold mines of California and Australia combined, could make me, with my matchless treasure by my side. No human beings could have greater reasons for joy and thankfulness than your dear Mamma and I had, by knowing that we have six such daughters and four such sons-in-law, each and every one of them a source of joy and pride. We have had on this occasion too the happiness of seeing three darling grandsons with us. We have received many very pretty presents, but none prettier or more to be valued than those that came from Berlin.

Your letter this morning, my dearest Nora, written so recently as the day before yesterday, made us very happy, and

most glad to receive so good an account of dear little Annie,
her little lock of hair which you sent me is in my pocket-
book.

God bless you both,
Your affectionate father,
LEONARD HORNER.

To his Daughter.

17, Queen's Road West, *5th July*, 1856.

MY DEAREST LEONORA,—We hope to join you at the place
of meeting in the Hartz. Where that is to be, the wise experienced heads at Berlin must determine. Süderode must
be a very small place, as it is only set down on a map of a
large scale. I am now busy *cramming* the geology of that
region. It will be full of interest to me, and I hope to find a
quiet pony to take me about, with a boy to hold it while I
descend to ply my hammer. There is much geological, and
probably picturesque variety very near Süderode, and the
course of the Wippra will afford many subjects for the pencil,
particularly Schloss Rammelsberg and the environs of Mägdes-
prung. A visit to the Brocken must be a special expedition,
and it would be a shame to be in the Hartzgebirge and not
visit that far famed mountain. About the 15th of September
we must proceed direct for England.

Love to you all from,
Your affectionate father.
LEONARD HORNER.

To his Daughter.

18, Rue de Montaigne, Paris, *July 25th*, 1856.

MY DEAREST KATHARINE,—We are enjoying our visit here.
It is impossible that more affectionate kindness could be shewn
than that which we receive from your dear aunts and from

Byrne. I have made the acquaintance of an able geologist and pleasing mannered man, M. Delesse, and I am to meet him at breakfast on Sunday at the lodgings of my friend Dr. Andrews of Belfast, who is here with his family for some months. I spent yesterday pleasantly and profitably in the Geological Museum at the Jardin des Plantes, and am going again to-day; I met there Mr. Charles D'Orbigny, an old acquaintance; he is the Assistant Professor of geology to M. Cordier. The latter is a remarkable man, for although eighty, he is as active and looks as fresh as a man of sixty. He is one of two remaining of the distinguished corps of scientific men who accompanied Buonaparte to Egypt in 1798; M. Jomard is the other, but he is very infirm.

This is certainly a most wonderfully fine and most interesting place. I think that I should enjoy staying here some weeks, were I to do so in perfect quiet, and with full leisure to go about everything calmly and deliberately.

Pray write *on Monday*, directing here, and be very full about yourself, and give me a good account, as I trust you will be able to do of Harry, darling Arthur, and the little pet Rosamond (the first time I have written her name).

Your affectionate father,
LEONARD HORNER.

To Sir Charles Bunbury.
Suderode, bei Gernrode, Preussen, <i>August 13th</i>, 1856.</div>

MY DEAR CHARLES,—I waited the arrival of the dear party from London to write to you, and that joyful event took place two hours ago. . . . We arrived here a week ago, and I found George, Leonora, Joanna, and the two dear boys, with Hermann ready to welcome us. Our ten days at Paris were spent most pleasantly, and I saw a good deal of Delesse, and something of Charles d'Orbigny. I spent a great part of my time in the Mineralogical Museum at the Jardin des Plantes. This is a very pretty place, and the walks among the wooded hills are quite charming. We have made one expedition where we passed into a narrow gorge of the range in which the river Bode flows, planked on both sides by lofty perpendicular cliffs and needles of granite of vast height. No donkey or mule is to be had here, which I might have availed myself of. I expect, however, to be able to see some of the great features, that are not very distant. But I am not discontented, I am reading German assiduously, and while I am in the country, will try to improve my acquaintance with the language as much as possible. Then, to say nothing of the happiness of so large a family party, we have endless beautiful walks. George and Leonora are in excellent health and spirits, and most happy with their dear child, as well they may be.

With our united affectionate love to you both.

I am, ever my dear Charles,

Most truly yours,

LEONARD HORNER.

———

To his Daughter.

Queen's Road, West, 9*th October*, 1856.

MY DEAREST MARY,—You may be very sure that the first thoughts of all of us this morning were of you. It was the rising of a bright sun of happiness, one that has never been clouded now for forty-eight years. Wherever you are, there is sunshine. God bless you, may you never see this day return without cheerfulness. It was a great pleasure to your dear Mamma to receive your letter from Munich of last Monday ; I read it to her before she was up and it cheered her. It is a great happiness that both you and Charles were so well and having so much enjoyment in the beautiful and interesting countries you have visited.

Your dear Mamma is still an object of great anxiety to me. I shall not go to Manchester until I see her thoroughly convalescent.

My kindest love to Charles.

Ever my dearest Mary's affectionate father,

LEONARD HORNER.

I do not think that I have expressed to you the great joy I have had in the elevation of my friend Tait to the See of London. I always predicted that he would be a Bishop, but never expected to live to see him one.

To his Daughter.

17, Queen's Road, West, November 30th, 1856.

MY DEAREST FRANCES,—You will readily believe that the engrossing subject of my thoughts at present is, and has been for three months and a half, the state of your dear mother. It has been a great suffering to me to see what she has suffered. I am willing to believe from all I see, but still more from all I am told, that before very long I shall see her restored to good health, but that would be such an inestimable blessing, that a fear constantly comes over me that it will not be realized. It is an unhappy tendency of my nature to dwell on possible evils, when there is anything approaching to a cause; and a perfect consciousness of the absence of sound philosophy in this, and still more the experience I have had so often of my anticipations not being realized, have not succeeded in effecting a cure. Patience in such a case is easily prescribed, both to her, and to those who are afflicted by seeing her suffer, but very difficult to practice. She never, during the whole time it has been my chief blessing to know

her, now in the fifty-third year, has had any illness to compare
with this, for which God be praised.

The stillness of our lives has given me much time for
reading. I have dined out five times in the last two months,
once at the Wedgwoods a month ago, expressly to meet my
old friend Fanny Allen, twice at the Philosophical Club, and
twice at the Geological Society Club, each time my mind
having been tranquilized by a temporary improvement in our
patient. My reading has been very various, Schiller's Life,
some of his plays, and a considerable proportion of geology.
Having found sometimes my mind not in a fit state for severe
attention, I. took up, after an interval of half a century, *Tom
Jones!* I read a volume and a half, and was so disgusted
with its coarseness that I closed it, never to open it again. I
sought relief and tranquil enjoyment in biography, and took
up Dugald Stewart's "Lives of Adam Smith, Robertson, and
Reid," and found both an ample store in that charming
volume. Your Charles tells me that he has been reading
Mackintosh's Dissertation on Ethical Philosophy. and that he
has been so charmed with it that he intends to take up some
other work on Moral Philosophy. Has he read Smith's
" Moral Sentiments ? " If not, he will derive much pleasure
from the work ; it would be a good introduction were he to
read Stewart's " Life of the Author." Your Uncle Horner
used to say, that Stewart's Preliminary Dissertation, in the
same work as that of Mackintosh, was, in his opinion, the
most finished of all his writings, and the most generally
interesting.

The last book I have read, has been Baden Powell's second
edition of his work on the "Unity and Plurality of Worlds,"
which has afforded me much pleasure and instruction. It is
written in the true spirit of the inductive philosophy. He
very ably refutes both Whewell and Brewster, although they
are opposed to each other. I felt not a little pride and

pleasure at seeing in one page the names of Galileo and Lyell
in juxta-position.

With affectionate love to your husband,

Believe me, my very dear Frances,

Your affectionate father,

LEONARD HORNER.

To his Wife.

53, Harley Street, 16*th December*, 1856.

After closing my letter to Joanna yesterday at the
Athenæum, I walked back here.

We had a pleasant, sociable party at the Moore's; the
family including Mrs. Moore and Graham, and John just
arrived from Scotland with James Spedding and Mrs.
Robinson. I sat between her and Julia. Charles told a good
story of Brougham's reply to Lord Crewe, who on some rare
occasion when he voted with Brougham, said to him, "We
are in the same boat to-day," to which Brougham said in
reply "Yes, but with very different sculls" (skulls).

Esmeade told another good reply; Phillips the barrister, a
great talker, said "How very odd that the hair on my head
continues brown, and my whiskers are grey." "No wonder
at all," said Brougham, "because you work your head so
little, and your jaws so much."

To-morrow is Geological Society day, and I mean to dine at
the Club.

To his Wife.

17, Queen's Road West, *December* 18*th*, 1856.

Your dear letter and that of Joanna, which I found at ten
o'clock last night when Charles and I returned from the
Geological Society, made me very happy, and sent me to bed
with an easy mind. You must not be "low spirited from an

anxiety to be stronger and not to be such a dependent creature," because that very anxiety retards the work of your getting stronger. I am not at all disheartened; I have long been satisfied that you must be treated as a hot-house plant all the winter and spring, and it will be my greatest pleasure and that of all your devoted children, to tend you and do everything in our power to strengthen you on the one hand, and on the other to give you all the amusement that your tranquil life is capable of; and I doubt not you will find many objects of interest.

I am glad that Joanna saw Mrs. Buckland. If she sees her again, give her my kind regards.

I left dear Charles and Mary after breakfast. They were ·
to go at twelve o'clock to the railway. I found everything prepared for me at No. 14, in the nicest and most luxurious way. What a beautiful spare room dearest Kate has made. All the children are very well and very cheerful, Leonard and Frankie delighted with dear Grandmamma's letters to them, which they showed me, and well they might be delighted with such charming letters, beautiful both in substance and in writing. As Kate justly remarked, how very few ladies of seventy ever write so clear, so firm a hand—another proof that your nerves are very strong. 1 am now writing in the study, and all the maids are well, and the quadrupeds and bird also.

My love to my dearest Joanna.

I am, my very dear Anne,

Your devoted and loving husband.

———

To his Wife.

17, Queen's Road West, *December* 20*th*, 1856.

My VERY DEAR WIFE,—To-morrow is the anniversary of that blessed day which gave birth to her who has been the chief object of my thoughts for more than fifty-two years.

Never did man possess a more devoted affectionate wife.
I do not remember that we ever before were separated on
your natal day, or that I ever saw you otherwise than in
perfect health until this sad year. But, praised be God for
it, you are greatly restored, and by His blessing I hope you
will very soon be perfectly so, and that I shall never again
see the 21st of December without you by my side in the
enjoyment of all earthly happiness.

Katharine and I went yesterday evening to Highgate School.
The examination was in the school room, which was quite
crowded with spectators. Our coming was not expected and
we were kindly greeted. They put me in the chair, so I had
to make a speech at the close. The children answered most
creditably, and Mr. Mumford arranged and conducted the
examination admirably. Harry went to the pit of the
Princess's Theatre and saw the "Midsummer Night's Dream,"
with which he was much pleased.

I enclose two charming letters to you from our dearest Nora
and George. I also send a letter I have had to-day from that
charming and wonderful old lady Mrs. Fletcher.

Harry and Katharine and the children are all well. By a
letter from Mary to Katharine they are enjoying themselves at
Bowood.

God bless you and protect you, my dearest Anne.

CHAPTER XII.

1857—1860.

To his Daughter.

17, Queen's Road West, *6th February*, 1857.

MY DEAREST LEONORA,—I am reading Friedrich Perthes
"Leben," and am deeply interested by it. He was a most
remarkable man, so wise, so prudent, so energetic, and so
modest, overcoming by his own talents and virtues all the dis-
advantages of his early life, and acquiring a just rank and
influence with the most eminent men in his country. There
are two interviews which Stein described, one at Frankfort,
the other at Stein's own house at Nassau, very interesting.
If you have not read it, I think you would derive as much
pleasure as we have done, for the translation has been read
by your Mamma and Katharine.

I am delighted at the near prospect of seeing you my
dearest Nora and your little darling.

I wish I could say that your dearest Mamma was quite
well. She is far from that, and a great sufferer, never twenty-
four hours without much pain. I thank God that I am
assured there is not danger to her precious life, but it is sad
to see the dear soul suffering so much.

God bless you dearest,

Your affectionate father,

LEONARD HORNER.

To his Wife.

London, *February* 21st, 1857.

Yesterday forenoon the Bishop of Rupert's Land called. I traced in his features those of the Academy boy.* I was much pleased with the interview, I liked his manner much, and he seemed and expressed himself gratified by the renewal of our acquaintance. He brought his sister with him. She has been with him in his remote diocese, and is going out again with him in August. They have a post through the United States once a month, but all other communication is restricted to once a year, when the Hudson's Bay Company's ship arrives at York Factory on Hudson's Bay. He expects to be in London in May, when I hope to see him again.

The Geological Society's dinner went off well. We had a very good speech from Owen, and also an excellent one from Lord Panmure on the importance of better educating the officers of the army in science, and describing what has been done and what is still doing towards that end, and highly approving of appointments being given by competitive examinations and not from patronage as heretofore. I was not at home till half past eleven.

God bless you my darling. My best love to Charles and Frances.

———

To his Wife.

London, *March* 3rd, 1857.

MY VERY DEAR ANNE,— . . . . On Sunday I was at home all the forenoon but went to St. James's, Piccadilly, at three, to hear the Bishop of London preach. I never was in that church before, and it is a handsome one, and the service

* David Anderson, who was at the Edinburgh Academy the same year as Archbishop Tait.

well performed in the manner of the olden time, simply and
decently, without parade of any kind.

We had Mr. Hochsetter at dinner, introduced by Haidinger;
he is going out in the Novara Austrian frigate, round the
world. The chief object of the voyage he says, seems to him
to be, that the Austrian flag as a maritime power may be
seen in places where it has been hitherto unknown. Yesterday
I dined at Owen's, we went together from the railway station,
and Broderip brought me back in his carriage. It was a
pleasant visit, the house is quite a *bijou*. To-day the Arthur
Herveys have been with us at luncheon, and I left them in
the Zoological Gardens to come home to write this letter.

Athenæum, March 5th. I will tell you a story, which that
amusing man Broderip has just been telling me. When
Soyer was cook at the Reform Club, he lost his wife, and in a
sad doleful strain, he told his sorrow to Sergeant Murphy,
and asked him if he could suggest an appropriate epitaph to a
monument to his " dear angel " which he was about to erect.
Upon which the Sergeant replied, that he thought he could
not inscribe anything more appropriate than *Soyez tranquille.*

I have just come from Owen's lecture, which was, as usual,
very good. Mary, Katharine and Joanna were there. I sat
next to Lord Arthur and Lady Arthur Hervey, and Lord
Bristol was with them.

Yesterday after dinner I read in Hume the last chapter of
the reign of Edward III, and the reign of Richard II, preparatory to reading Shakespeare's Richard II, and why did I do
so ? because Mrs. Kean told me that the next great dramatic
exhibition her husband is to give, is, to bring out with great
effect that play. To-day I dine at the Philosophical Club,
to-morrow at Harry's, and next day I trust to see your dear
face opposite to me at our own table.

Give my love to my dearest Frances and her husband.

Your affectionate husband,
LEONARD HORNER.

To his Daughter.

Folkestone, *August 27th*, 1857.

MY DEAREST FRANCES,—I enclose a most interesting geological letter from Charles Lyell, which your husband especially will read with much interest. It is quite delightful to see with what youthful ardour, and with how candid a spirit, Lyell pursues his scientific inquiries. I have a constant dread of his eagerness combined with his short sight, leading him into such dangerous ground among the rocks, that a serious accident may happen; which God defend him from. He will have intense pleasure in visiting the Glaciers of Zermatt, so well described by James Forbes   I was very glad to read dearest Mary's letter to you from Trogen.

We have glorious weather. The sea was this morning, and now is, as smooth as glass, the motion at the water's edge is scarcely perceptible. Dear Mamma is well to-day, although she had not a good night.

Affectionately yours,

LEONARD HORNER.

To his Daughter.

Folkestone, *September 5th*, 1857.

MY DEAREST MARY,—I can continue a good report of your dear Mamma, and confidently expect to take her home this day week decidedly improved, and such is her own opinion. I shall miss my walk on the cliff before breakfast; I might substitute Primrose Hill for a week, but there is no hope at Manchester. I never went there with so much reluctance, arising no doubt from the disappointment of my hope that an improved Superannuation Act would have passed, so as to enable me to retire. But perhaps I am neither reasonable nor a good philosopher, in not considering that the right view to take, is that of thankfulness, that at the age of nearly 73, I am

able to go through the work without the consciousness of
any mental unfitness for the work, and with a fair allowance
of bodily strength.   If my mind continues at ease about your
dear Mamma, and if you are all preserved to me in good
health, all other things ought to be viewed as trifles, however
they may be momentarily irksome.   Neither do I feel any
abatement of my interest in science, and almost every day
since I left home I have been working on my Egyptian paper.
It has been satisfactory to me, to see that these researches
have excited some attention among persons whose attention is
valuable, for Dr. Lloyd in his address last week at the
meeting of the British Association at Dublin, as President,
noticed them.

I anticipate great enjoyment to both of you from your visit
to Italy, and I do not doubt that Charles will see much in the
phænomena of Vesuvius and Etna, that he will turn to good
account, and may learn much that will have important
bearing on his Madeira, Teneriffe, and Canary Island papers,
to all of which I look forward with great interest, especially
after the foretaste of that he has given in the last edition of
the Manual.   His letters from Bienne and from Zürich, have
been read by Charles Bunbury, Hooker and Darwin.   Darwin
says, " What a wonderful man Lyell is ; he has steam enough
to make half a dozen ordinary geologists.   The leaves in the
lower cretaceous beds are grand, and I was particularly
interested in hearing about the great difference in species as
to the plants and insects compared with the shells in the
Swiss Tertiaries."   Poor fellow he gives but an indifferent
account of himself.   He says, " My health has been below my
poor par for some time, and I fear, what 1 never used to
believe, that all mental work of any kind, kills my stomach.
Nevertheless, I can, and will continue at work on my book on
the origin of Species, which makes slow but steady progress."

The meeting at Dublin appears to have gone off most

successfully. I have read Dr. Lloyd's address in the Athenæum which is most excellent. I have not as yet seen any account of the work in the Sections—among the geologists present I have only seen the names of Portlock, Daubeny, Rogers and Phillips—of course Griffith would be there. I fear that Sedgwick was not well enough to go. Murchison is I hear in Germany. Give my love to Charles, admonish him to be cautious not to run into any danger in his explorations, to remember that he is not so young as he was, and that he is still short-sighted. God bless you both.

Your affectionate father,

LEONARD HORNER.

To his Daughter.

Manchester, *September 24th*, 1857.

MY DEAREST KATHARINE,—I left home on Tuesday morning and arrived at Manchester by half-past four. Having written to the landlord of the Albion as an old customer, he had kept a room for me, although he had turned away fifty parties, such is the concourse of people still to visit the Exhibition, suddenly augmented by the announcement that it is to be closed on the 15th of October. It was very ominous against my finding good lodgings, but I set out the same afternoon, and had not proceeded far, when I found the rooms from which I now write. My chief object of course was to seek for comforts for your dear Mamma, and I think she will be quite as well off as she was at Folkestone, so far as indoor comforts are concerned.

I do not expect that they will come before Monday, and so I must do the best I can in my solitude, which is very uncongenial to my nature. I had a letter from your Mamma this morning, giving a good account of herself, so I trust that she will in no degree suffer from coming here; I have no reason to fear that she will do so, but the last year has made

me tremblingly alive to all that can in the least degree
endanger so precious a treasure. She tells me that you
remain at Sheilhill until the 7th, I am not surprised that
you and Harry should have been glad to prolong your stay,
surrounded as you are by so much kindness and affection.
I long to hear the music of the voices of the dear children.

So soon as I had settled about the lodgings, I went to the
Exhibition. It fully deserves all the praise I have heard
bestowed upon it. I shall go as often as I can before it closes,
and am glad your dear Mamma and Joanna will see it.

On the dreadful subject of India which must occupy so much
of your thoughts, I cannot venture to touch. We have as yet
heard of no friends being murdered, but to-day I heard that
our acquaintances the Miss Ewarts here, had their brother,
his wife, and three children in Cawnpore, when the massacre
took place, and they were included.

Let me hear from you soon my dearest Kate, and the more
you tell me of your babes and their sayings and doings, the
better. Love to Harry, and my kindest regards to Tom and
the sisterhood.

Your affectionate father,

LEONARD HORNER.

To his Daughter.

Manchester, *October 9th*, 1857.

MY DEAREST MARY,—You may be very sure that we have
thought and talked much of you this day, your dear Mamma
and I blessing the day that gave us such a treasure. We
have had the happiness too this day of reading your letters to
Susan and to Sophy from Pisa. Your having been in the
room where your dear Uncle died, must have been very
interesting to you, and will leave an impression never to be
forgotten. You were quite right in your conjecture that the
room behind was where I slept. Every event of that sad, sad

day is as fresh in my memory as if they had occurred yesterday. I do not wonder at your admiration of Pisa. I hope to hear that you have seen the monument at Leghorn. Charles's letter finished at Pisa I have read with much interest, and have forwarded it to Charles Bunbury.

I have written to Ehrenberg to ask him if he has examined the *flysch*, and whether he has found animal remains in it, also if he has ever found any in dolomite. The latter is not probable, if dolomite be metamorphic, by heat, for in that case the organic remains would be melted with the matrix.

To-day you were to sail from Leghorn for Messina, so it will be long before we can hear of your arrival there, which I fervently trust may take place speedily, and without any suffering or mishap of any kind. Charles has judged wisely in going first to Etna to prevent any obstacle from snow. I have got notice to-day of the first meeting of the season of the Philosophical Club on the 15th. I shall not be there, but hope to be at the next meeting on the 19th of November.

God bless and preserve you darling Mary, and with best love to your kind, good husband.

I remain, your affectionate father,

LEONARD HORNER.

To Sir Charles Bunbury.

Manchester, *October* 18*th*, 1857.

MY DEAR BUNBURY,—I was very glad to see the letter of Lyell which you were good enough to send me and which I now return. I think it would interest Joseph Hooker very much, and considering the opinions he has published of the limited inferences to be drawn from fossil botany, and the deference paid to his authority, I think he ought to see what Lyell says of his now having "more faith in fossil botany, and more expectations of the services it is destined to afford us."

My before breakfast book has been "Cicero's Letters to

Atticus," with the Abbé Mongault's commentary. I have read five volumes of them, and they are certainly most interesting, although very often very difficult to understand from the brevity of the style, the allusions by *demi mots*. But they give a most graphic picture of the time, and shew what a set of scoundrels there then were among the leading public men. They too often shew the weak side of Cicero's character. I have another volume of them still to read. My evening book, *with good print for candle light,* has been Arnold's Roman History, which I had not read before. Not only is it a clear, lucid narrative, but it abounds with the noble, generous and highly virtuous sentiments which characterized that great and good man. You see that I am not occupied with recent literature. My reading, and still more my remembering powers, are extremely limited, and I have long since made up my mind to be extensively ignorant, in order to avoid the distraction of miscellaneous reading of the literature and science of the day; but there is field enough within my limited range to afford me untiring occupation and pleasure.

My love to my dear Fan,

Ever affectionately yours,

LEONARD HORNER.

From Charles J. F. Bunbury, Esq.

Mildenhall, *October 22nd,* 1857.

MY DEAR MR. HORNER,—I thank you much for your interesting letter of the 18th. I am much interested by the account you give of your Egyptian researches, and very glad that they are in such a state of forwardness. The great mass of facts of a very novel kind which you have collected, and arranged, will of itself be a most important addition to our stock of knowledge; and I feel satisfied, that whatever deductions you may see reason to draw from them will be

most carefully considered and worthy of the utmost attention.
I shall be delighted to see the result of your labours.

I have not read anything of Cicero's letters since I was at
college, and then but a small portion of them ; but I believe they
are very interesting. Middleton has made great use of them in
his Life of Cicero, which I read with great delight while I was
detained in Edinburgh by Fanny's illness. I remember that
one of the things which most struck me in reading that
book, as coming out in the strongest light from all the facts
related, was the excessive corruption and villainy of the
judicial body at Rome in the latter days of the republic, and
there can hardly, I think, be a worse vice in the internal state
of a country. That, at least, is an evil from which we in
England have for a long time been very free.

I quite agree with you in liking and admiring Arnold's
Roman History. I think he is in point of style one of the
very best of our modern writers, and the moral tone of his
work is delightful. It is not his fault that there is much that
is heavy in the first and second volumes; it is inevitably
tedious work to grope for the scattered grains of historical
truth amidst such accumulations of romance and error and
confusion ; conjectural history has neither the charm of
romance nor that of exact knowledge ; but when he comes to
the war with Pyrrhus, to the Punic wars, and above all to the
second Punic war, Arnold makes us full amends ; his third
volume is one of the most interesting historical narratives I
have ever read. It is a great loss to the world that he did
not live to complete his work. I am at present reading the
Life of Sir Thomas Munro. Several years ago I read more
about India than perhaps usually enters into the studies of
those who are not in any way connected with the country ; and
now I have taken up the subject again, wishing to fill up the
gaps in my knowledge. The book I am engaged upon is in
great part very dry,—long details about the revenue of

particular districts; but I find also much that is worth
remembering.   Munro was a very remarkable and eminent
man, with wonderful powers of work, great sagacity, great
determination, and very high principles.   He was one of the
men fitted to gain and to govern an empire; we need not look
far now-a-days for men qualified to lose an empire.

With much love to Mrs. Horner and the sisterhood,

I am, ever your affectionate son-in-law,

C. J. F. BUNBURY.

To Sir Charles Lyell.

17, Queen's Road West, *October 27th,* 1857.

MY DEAR CHARLES,—It is a great disappointment to us
that we have had no direct letter from you or Mary since you
left Florence.   Your arrival at Naples is in the *Times* of
yesterday, in a letter from their correspondent there, dated the
15th.   I hope it is true that both Vesuvius and Etna are in
activity.   Is it not rare that both should be so at the same
time?

The November number of the Journal will be out on
Magazine day.   Falconer has been very active in getting
his paper forward.   Jones tells me that Falconer has
been into Norfolk and has seen every collection of prebos-
cidean remains of which he could hear, and says that the
specimens sent to London for Owen's examination, were as
nothing compared with the perfect state of numerous specimens
he has seen, and that they partly confirm, partly oppose his
former opinions.

Huxley and Tyndall were on the Alps this autumn, working
on the Glaciers.   Huxley has published a paper in the
Philosophical Magazine describing the experiments he undertook on the Mer de Glace, Brenner, &c., on the internal
structure of Glacier ice, and he differs entirely from the
descriptions of Agassiz: indeed his paper is avowedly for the

purpose of shewing Agassiz to be wrong. He distinguishes *superficial* from *deep ice*, and shews that while the former is porous and full of fissures, the latter, even no deeper than a foot from the surface, is as compact and impervious to filtration as flint. He says that Agassiz did not make the distinction, but he also differs from him as to the existence and form of the globules, and that he makes that which is water *air*, and that which is air *water*. He does not even remotely touch upon the Forbes theory of plasticity, and consequently says not a word whether his discoveries have any bearing upon the motion of Glaciers.

You will doubtless have heard that the French Government offered to place Agassiz in the chair of Alcide d'Orbigny, and that he has declined the offer.

Love to my dearest Mary,

Your affectionate,

LEONARD HORNER.

To his Daughter.

17, Queen's Road, *November* 15*th*, 1857.

MY DEAREST FRANCES,—I send you a charming letter from Mary, from Catania, which came joyfully yesterday on dear Charles's birthday, and it was the more welcome that we had been so long of hearing direct from Mary. I rejoice to see not only that Charles has learned much by his visit to Etna, but that he has detected the unworthy proceeding of Elie de Beaumont in keeping out of sight cases of lava streams that had flowed at a high angle, and which he could not fail to have seen, as he and Dufresnoy described Etna so minutely, because they were opposed to a theory he was trying to support. There are thus you see party feelings, and party dishonesty where they ought never to be admitted. But I suppose it is a rare quality, the possession of that candour, which would be as ready to admit *a fact* against as in favour

of a preconceived opinion. I believe that the cultivation of
such a frame of mind is as necessary in science, as in politics
or religion.

I am very happy to see that dear Mary has found so many
of the inhabitants of Catania, attentive and hospitable to
her, and that she was able to see some parts of the celebrated
mountain. I shall be glad to hear of their being at Rome,
out of the dominions of Bomba.

We shall be delighted, you may be sure, to have you and
Charles at Christmas, and we hope to see much of you before
that time. We dined yesterday at No 14, to celebrate Charles
Lyell's sixtieth birthday. Leonard supremely happy, having for
the first time put on the *toga puerilis*, the crowning charm of
which were two pockets in the trousers, into which Grandmamma and Grandpapa dropped a shilling. All the four are
looking very well : Leonard is developed as much in mind I
think as he is in stature. There is a good prospect through
the kindness of Pillans, of their having an excellent daily
tutor for Leonard and Frankie, which will be a great relief to
Katharine.

With love to Charles, I am, my dearest Frances,
                    Your affectionate father,
                              LEONARD HORNER.

——————

To his Daughter.

Manchester, *April 4th*, 1858.

MY DEAREST MARY,—Mr. and Mrs. Ticknor's letters this
morning are very interesting like all they write. His opinion
of Buckle is great praise, for there are few better judges I
should think. I see that the book has made the same
impression on him that it has on myself, that with all its
faults, it is the most remarkable work that has appeared for a
very long time. Thick volume as it is, I look forward to

soon reading it a second time. I long much to see a fair review of it, by some one competent to such an undertaking, and there are very few of known authority able for such a task.

The India Bill of Disraeli, appeared to me from his statement, preferable to that announced by Palmerston, but I must wait to hear what wiser heads think of them. It will be and ought to be thoroughly sifted, and how much will remain in the sieve after the operation no one can predict. There never was a more important measure before Parliament.

Have you seen the third report of the Civil Service Commissioners? Sir E. Ryan sent me one, and I have read it with great interest. It completely answers all the ignorant objections made to their system of examination. One thing strikes me very much, and I must talk to Ryan about it : except for a single sentence, mentioning that a Mr. Longmore is an Examiner at Edinburgh, you would not suppose both from the report, and from the voluminous appendices, that there existed such a part of the United Kingdom as Scotland. Not the least account seems to have been taken of the different systems of instruction in Scotland from England. All the examiners in London are either Oxford or Cambridge men, so that a man coming before them who has been educated in Scotland, comes before Examiners generally imbued with a prejudice against Scottish schools and universities, and if he was ever so learned, if he made a false quantity, God help him.

I am rejoiced to hear that our dearest Susan looks better, and thinks herself better. How rejoiced I shall be when that destroyer of her health, Colletta,* is banished from her sight.

Love to her and Charles,

Your affectionate father,

LEONARD HORNER.

* His daughter Susan was engaged in writing a concluding Chapter to Colletta's History of Naples, which she had translated.

To his Daughter.

Mildenhall, *April 18th,* 1858.

MY DEAREST MARY,—I send you a communication I have this day received from Dr. Murray, the Secretary of the School of Arts, which you will readily believe has been very gratifying to me, and I am sure you will read it with pleasure.

I have restricted my prize to working mechanics, as it was for them that the school was instituted, and there is this additional reason, that many of the students are of a different class, who have more time for private study than a youth working at a handicraft trade can command, and therefore the competition, if open to all the students, would be unfair.

Next year the prize will be given in the Chemistry Class, next in that of Natural Philosophy, and then the Mathematics again in rotation, the three *staple* branches of instruction.

This competitive examination shows the substantial instruction which the School affords, and the diligence of the students. The legacies, especially that from an old pupil, are very gratifying, and a proof of that public interest in the school, which is likely to secure its permanence. Not only is there no debt, but the building, or rather house, is our own, and from two shops, yields about £100 a year of rent. But more money is wanted to pay the teachers better, and so secure able men.

Farewell my very dear Mary,
Your affectionate father,
LEONARD HORNER.

To his Wife.

63, Marine Parade, Brighton, *June 6th,* 1858.

We have just returned (seven p.m.) from a delightful drive to the Devil's Dyke on the north escarpment of the South Chalk Downs, looking over the extensive Weald of Sussex.

We had two carriages conveying the whole party except Rosamond. It has been a most delightful excursion. This trip to Brighton has been a most happy one for me : it is impossible for me to feel, far less describe, sufficiently, the affectionate kindness I receive from all, or the intense pleasure I derive from these delightful children. The only deduction is that the exuberant spirits of the three boys keep me in constant dread of their running into danger, especially Frankie. It would have been worth coming here to see what I did before dinner. How often I have wished, dearest, that you could have been with me to have had an equal share in all these joys.

God bless you darling, love to Susan and Lady Bell.

To his Daughter.

Manchester, *March* 28*th*, 1859.

MY DEAREST MARY,—I am glad to hear of your intended trip to Holland. I am always rejoiced at any cessation of Charles from work, and neither the country nor the season will be favourable to much field work. I do not know what Museum beyond that of Leyden is likely to occupy much of your time. I take it for granted that you will see Count von Randwyck often, and I hope that his mother is still alive and that she will be able to see you. It is now twenty-nine years nearly, since we made her acquaintance, and I think she could not have been less than fifty years old then. I hope too that you will see at Amsterdam my old friend Samuel Labouchère, now past eighty, but still active, as his son lately told me, and Young says that his hand-writing, which they often see, remains the same as ever.

I have finished reading your dear and lamented friend Mr. Prescott's last volume, which I greatly prize as his gift. I have been greatly interested by it. It is fearful to think of

the horrors committed upon and by the Moriscos. His account
of Don John of Austria is very striking, particularly of his
early life; but the humanity for which he gives him credit on
many occasions, seems entirely to have left him when he took
the command in the Low Countries, according to what Motley
has told us of his deeds there. Mr. Prescott gives a more
favourable view of Phillip's character than I ever supposed
could have been given of so cold-blooded a monster as Motley
makes him. I must read again the first two volumes of his
reign, but these do not come down to the time when his
cruelties in the Netherlands first began. If I were younger I
would set about learning Spanish to enable me to read
Mendoza's " Guerra di Granada," and some of the other
authors he quotes. You must shew me what Mr. Ticknor has
said of Mendoza.

Have you heard whether Mr. Prescott has left his materials
for the continuation of his history of Philip in such a state
that they can be published. It is sad that the most interesting, at least the most exciting part of his reign, the affairs
of the Netherlands, were not all to come from his hand.

Ever, my dearest Mary,

Your affectionate father,

LEONARD HORNER.

To his Daughter.

17, Queen's Road West, *June* 16*th*, 1859.

MY DEAREST FRANCES,—I wish I could see your roses, and
these with the tranquillity of the white room, and some opportunity of reading, would be a great contrast to our life here.
I scarcely ever get an hour's quiet reading; I must say however, that I do spend a good deal of time, all I can spare, at
the Geological Society, in the herculean task I have undertaken, to get our museum out of its discreditable disorder.
The task would be easier if I could find some one to help me,

but that I have found very difficult. But I am making some progress by sticking to the work.

The Italian cause is prospering gloriously, so far as the defeats of their Austrian oppressors have yet promised. *Louis Blanc* told Pulsky that he is confident that Napoleon will not attempt to make any conquest in Italy, *because* it would be against him in France, but that to be held up as the *Liberateur de l'Italie* will immensely add to his popularity. This from Louis Blanc is remarkable. My best love to Charles.

> Your affectionate father,
>
> LEONARD HORNER.

To his Daughter.

Edinburgh, 28*th* *July*, 1859.

MY DEAREST JOANNA,—I wrote to Hekekyan Bey in April, sending a part of the *Athenæum* containing Samuel Sharpe's objection, that when the pedestal of Ramesses II. was erected, the inundations of the Nile must have been prevented from overflowing the site by embankments around Memphis. Some weeks ago I sent to him the number of the *Quarterly Review*, and told him that I abstained from all comments on either criticism, wishing to have his unbiassed opinion on the objections. I also requested Mr. Harris of Alexandria, who has lived more than thirty years in Egypt, and is acknowledged as an antiquarian of undoubted authority, to give me his opinion, particularly calling his attention to the statement that bricks were unknown in Egypt before the Romans. I have not had as yet any reply from either. If, as is probable, there was an embankment, Sharpe's objection is good; but as to the amount to which it will affect the rate of secular increase, nothing can be said until we know at what period the Nile inundation again overflowed the site, that it had for ages overflowed that ground is proved, as the pedestal of Ramesses rests (with the intervention of a foundation of

carried sand) on Nile mud, and that mud was penetrated to
the depth of thirty feet below the pedestal. Will you let
Charles know all this. He will be pleased to hear that the
Calvinistic clergy of Scotland are throwing overboard Archbishop Usher's Bible Chronology of the Creation. Mrs.
Cunningham told us two days ago that *she heard* Dr. Guthrie
say from *the pulpit* that men of accurate science had shewn
that the globe we inhabit is of vast antiquity, that 20,000
years were in all probability but a part of the age. Mrs.
Cunningham said that a new light had broken in upon
her, that many of the congregation were dumbfounded,
and said that he must have lost his senses; I hope he will
state this belief in print. It will be somewhat remarkable if
the Scotch Presbyterians shew an enlightenment on such
subjects in advance of the *so-called* more enlightened clergy
of the Church of England.

Your affectionate father,

LEONARD HORNER.

To his Daughter.

Drumkilbo, *August 7th*, 1859.

MY DEAREST JOANNA,—I have had a letter from Mr. Harris.
He says that " it may be true that no burnt bricks have been
found in buildings before the Roman dominion, the reason
being that, as at the present day, unburnt bricks served the
purpose of housebuilding in the interior better than any other
material. Such bricks still remain in all ancient ruins, for
in this country, time lays but a very gentle hand on materials
the most perishable. The Reviewer cannot mean that the
Egyptians did not make use of burnt clay for other purposes,
or he would be wrong, for the museums abound with proofs to
the contrary, and I have before me a burnt brick having on it
an inscription in hieratic, date unknown."

Now there cannot be a doubt that the Reviewer intended

his readers to believe that burnt bricks or pottery did not exist, or were unknown in Egypt, before the Roman dominion. The review was in part at least, written in the British Museum, and it will not be difficult for me, I apprehend, to point to pottery in that collection more than a thousand years old before the existence of Rome.

With regard to Mr. Sharpe's objection, he only says that it is certain Memphis was surrounded by a dyke in the time of Herodotus. On this point, 1 wait the reply of Hekekyan Bey.

Affectionately yours,

LEONARD HORNER.

To his Daughter.

Drumkilbo, Perthshire, *August 7th*, 1859.

A beautiful day, my dearest Susan, we have had a large share of good weather ever since we left home.

We intend to leave this place, where we have had great kindness and hospitality and much pleasure, to-morrow morning.

On Friday we drove to Airlie Castle; the well cultivated fields of Strathmore and the undulating country with some gentlemen's seats, and many capital farmhouses, make the drive interesting, and when we come upon Airlie Den, with the Isla winding and sometimes roaring through the steep wooded banks, the interest increases. A vast wall and portions of a tower shew that it must have been a castle of great strength. Yesterday we drove to a lowland comfortable house, with pretty grounds and gardens : Kinloch, the proprietor, being of that Ilk. Woods and evergreens growing most luxuriantly, and a garden well-stocked with beautiful flowers amid the substantial requisites for the kitchen. The evergreens both there and here are very fine.

I have been reading a very interesting book by Hugh

Miller.—" The Cruise of the Betsy among the Hebrides." The Betsy was a yacht provided for the minister of the Parish of Eigg, after he had gone over to the Free Kirk,—land to build a church for him on the island having been refused by the landlord; one of those disgraceful acts of cruel intolerance which occurred at the disruption. It was called the parish of the small Isles, including Eigg, Rum, Muck, Canna and some others, the Manse being in Eigg. The same book gives not only a curious account of the hard service of such a minister, and very fine descriptions of the grand scenery, but many valuable geological facts, and wonderful proofs of the vast changes the land and adjoining seas have undergone in countless ages, up to a comparatively recent period.

Your affectionate father,

LEONARD HORNER.

To his Daughter.

Sandown, *September* 18*th*, 1859.

MY DEAREST MARY,—I have to thank you for a copy of the Prince's address which I received this morning. It does the Prince infinite honour, and I trust that it will have the effect of raising him in public opinion to a much greater extent than, as it appears to me, has yet been the case at all commensurate to his great merits.

Germany may well be proud that the education she affords on so extensively comprehensive a scale, has sent forth such an example of the training even her Princes may receive. I trust that this address will make such an impression on our Statesmen, as will greatly increase the influence of the Prince for the advancement of public scientific objects. How well he puts the merit of the scientific *bore*, of his adaptation " for the ends for which nature intended him." His address fully justifies the selection of the Prince as President, and makes me retract all I said in a former letter on this score. I am glad to learn that you are to be present at the *banquet* at Balmoral next Thursday, a royal luncheon could not be designated by a meaner term.   Remember us most kindly to the Clarks.

I expected to have found here some objects of geological interest, within my reach, but in this I have been disappointed. The chalk of Culver cliff is too far off, and the greensand cliffs and small patch of wealden are, as far as we have been able to discover, entirely without fossils, so that there is not even the small amusement of picking them out.

My love to Charles,

Your affectionate father,

LEONARD HORNER.

To his Daughter.

London, *July 7th,* 1860.

MY DEAREST LEONORA,—I hope soon to hear the day you have fixed for your departure for your country quarters.   It must be a great pleasure to you and Joanna to have Matilda Oersted with you.

My visit to Oxford turned out very satisfactory.   Professor Sedgwick was in the carriage with me, he was in good spirits, and talked all the way, often very amusingly.   On my arrival I went to the Reception room at the Town Hall, and there learned that rooms were provided for me in Magdalen College. Next morning, I went to the Council of the Association, when preliminaries were settled.   Sedgwick was chosen President, Charles Lyell, General Portlock and I were chosen as Vice-Presidents.   I have always preferred at those meetings to stick by one Section, and of course that of geology, and there was the additional reason this time, that being a Vice President it was proper that I should support the chair.   We had a good deal of interesting matter before us all the week.   I had time, after the close of the Section each day, to walk before dinner about the beautiful town, which I had not seen for many

years.   The colleges are most of them most picturesque
objects, and their gardens and particularly the smooth shaven
velvet turf, beautiful.   I dined at Dr. Jacobson's, a Canon of
Christ Church and Professor of Divinity, whom I knew when
we were at Bonn ; as he was a friend of John Sterling, and
came with him.   Next day I dined at Dr. Daubeny's, Professor
of Botany, who has a charming house in the Botanic garden.
On Saturday, I dined in Brazenose College with the Rev. Mr.
Symonds, a geological friend, a right liberal man, Rector of a
parish in Herefordshire, and a great friend of the Dean of
Hereford.   On Sunday, Sir Richard Griffith of Dublin, Dr.
Andrews of Belfast, and I dined together at the Mitre Tavern,
and we had a charming walk on a fine evening in Christ
Church meadows.   Next day I dined at the Vice Chancellor's,
Dr. Jeune, in Pembroke College, of which he is the chief, and
met a large and pleasant party.   I sat next to Madame Mohl,
who was staying in the house.   Every evening there was a
general assembly in the new Museum.   On Sunday I attended
the service in the University Church, St. Mary's, where in the
morning, I heard a most admirable sermon from Dr. Temple,
the head master of Rugby, which is to be printed, and I will
take care to bring a copy with me.   Lord Wrottesley's opening
address as President was very good.   We had Dr. Geinitz of
Dresden, a first-rate geologist, and a pleasant man.   I saw
very little of Charles and Mary, as they were living in another
quarter, but we met each day.   I have thus given you a full
account of my visit to Oxford, from which you will see that I
passed my time very pleasantly.

Ever dearest Leonora,

Your affectionate father,

LEONARD HORNER.

To his Daughter.

London, 11*th August*, 1860.

MY DEAREST FRANCES,—I have had the gratification to receive the enclosed. The original is handsomely framed and ornaments our dining-room.*

I condole with you upon the deplorable weather. It poured in torrents all night. It has been particularly unfortunate that so much of the enjoyment of Barton to your visitors of all ages has been so sadly interfered with. Give my kind love to Charles, Katharine, Harry and the dear children.

Your affectionate father,

LEONARD HORNER.

To Emily, Lady Bunbury.

Berlin, *September* 16*th*, 1860.

DEAR LADY BUNBURY,—I thank you most sincerely for the kindness which led you to write to me after you had read the paper† which my good wife had got printed to send to those who she thought would be sure to have a kind feeling towards me. I should have been very insensible indeed had I not been much gratified by such an appreciation of my endeavours to give full effect to the benevolent and wise interference of legislation on behalf of the oppressed factory population. It was a great experiment, and twenty-five years of testing the soundness of the measure, have shewn that moderation in the hours of work, are perfectly compatible with an ample remuneration to the capitalist. Sound, however, as the legislation has been in the case of the factories, the application of the principle to other employments would require great forethought and a perfect

---

* A testimonial from the working men in the factories, on Mr. Horner retiring from office. Now in the possession of his grandson, Leonard Lyell, at Kinnordy, N.B.

† Copy of the factory delegates testimonial.

acquaintance with the nature of the employment proposed to
be interfered with.   There is also this great difficulty, that
all legislation would speedily become a dead letter, without a
similar system of inspection to secure the observance of the
law, and that implies a very large expense.

We had a very happy family meeting at Rudolstadt in
Thuringia, which lasted some weeks.   The Pertz family
from Berlin, Sir Charles Lyell and Mary, Mrs. Horner, Susan,
Joanna, and myself.   The last four are now here, and propose
to remain some time.   We spent a week at Dresden, where
Susan had the greatest enjoyment daily in the wonderfully
fine picture gallery.

All here desire to be most kindly remembered to you.   I
beg you to give my kind regards to Colonel and Mrs. Bunbury,
and believe me,

Faithfully yours,

LEONARD HORNER

———

To his Daughter.

Berlin, *October 5th*, 1860.

MY VERY DEAR FRANCES,—I do not think that any letter
has passed between us since our visit to Barton, nearly three
months ago, which I often think of with much pleasure.

Our stay at beautiful Rudolstadt was most agreeable, and
we all had much enjoyment in our week at Dresden, Susan
supremely happy in the gallery, your Mamma much interested
by it and other sights, and I was both instructed and amused
by the geological Museum, and the conversation of Professor
Geinitz, who devoted much of his time to .me, including an
interesting geological excursion to the neighbourhood of
Meissen one day, and to the Plauensche Grund another.   But
this visit to Dresden will be especially marked by Susan's
purchase of Winkler's beautiful copy of the Madonna di San

Sisto, which now adorns the front drawing-room in Harley Street.

We have now been three weeks here, with much happiness. It is delightful to see our dear Leonora so perfectly happy with her excellent kind husband, and her dear engaging children. George went ten days ago to Munich, to be present at a meeting of historians, annually invited by the present king, who personally takes part in their proceedings. We expect him back on Sunday.

The Professors have hardly yet returned from their holidays. I have seen Gustav Rose, the Professor of Geology, and have passed some time with him in his museum.

Berlin has all the appearance of a thriving place, and I am rejoiced to hear that politically it has made considerable advancement during the regency, which has now existed nearly three years. One great step in advancement is the freedom of the press, the newspapers, three of which I see daily, seem to me to express opinions as freely as our own. There is no censorship, the chief of police has the power to stop the issue of a newspaper for a day, but the ground of his exercise of power, must within twenty-four hours be decided upon by a judge. A law similar to our Habeas Corpus Act is in full operation; the power of the Landsräthe, justices of peace, so shamefully abused under the last Government of the king, is rigorously controuled in the matter of the elections and the voting is brought near to the voter's dwelling-place, instead of being often fifty miles distant. All religious creeds are practically equal before the law, and are open to all employments, and great improvements in the primary schools and condition of the schoolmasters are in progress. The liberal measures of the Government were so obstinately opposed by the upper chamber that the Regent has just resorted to the extreme measure of creating one batch of eighteen peers for life; so that the house of peers will probably be brought to behave better. The feeling of the country is strongly in favour of the Italian cause, and the course of the government, in respect of Austria, is said to be, that Austria must settle her Italian affairs with the Italians. I am struck with the beauty of the domestic architecture in the many new streets that have lately been built. There are many architects, and all are said to be pupils of the late Schinkel, the architect of the new Museum. But there has been a great increase in house-rent. With all the external beauty there is much to do in the way of internal comfort. The plan of *flats* of apartments continues, and below the ground floor of a magnificent palace-like house, you have cellars occupied by small tradespeople, and behind, in a court, a number of small houses, occupied by tailors, shoemakers, carpenters, &c., with their accompaniments of noise and dirty children. The horrible smelling gutters in the streets continue, and they say that a cure for this is impossible, because of the want of fall ; from the end of the Friedrich's Strasse to the river, a distance of more than an English mile, the fall is only fifteen feet. But with all this display of handsome houses, in this part of the town, the following astounding fact shows how little wealth there is in the place. The population is 450,000 and the Income Tax is not levied upon any one whose income is less than 1,000 thalers (£150) a year. Not more than 10,000 are found to be liable to the tax. This fact I had from Professor Dubois Raymond, who had it himself from one of the Commissioners of the tax.

We have not yet been to any play, for no one has been acted since we came that has attracted us. Your Mamma, Susan, Joanna, and I heard Don Giovanni in German words ; Mamma and Joanna, Figaro's Hochzeit. I do not think that I shall ever go to an opera again unless it be the *Zaüber flötte*.

We shall stay over Annie's birthday on the 15th, probably

two days ; and with the exception of a stay of two or three days at Bonn, we shall go direct home by Calais.

Your Mamma had a most agreeable and entertaining letter yesterday from Lady Bell, who from what she says, will I suppose, be with you when this reaches you. Give her our best love. Tell her that I had read in the newspaper with much regret the death of my schoolfellow, Campbell of Craigie. It is another warning for me to be ready. The same post brought one to me from dearest Katharine, who I suppose is now in Queen's Road. It is very unfortunate that during both their visits the weather was so bad. I was glad to learn that Charles, when you were at Lowestoft, had had an opportunity of becoming better acquainted with dear Leonard, and had discovered how intelligent he is, and how charming his disposition and manners are. If it should please God to prolong his life, he will be a comfort to his parents, and be a favourite with all who know him.

When at Rudolstadt, I met in a bookseller's shop, a small book lately published—*Geist aus Seneca's Werken uber die höchster Angelegenheiten des Lebens*, drawn up by the Pfarrer of the neighbouring village of Kirchhasel. I have read the greater part of it, and am much pleased with it. It contains much that you may turn to account in both your schools, and I have bought a copy for you here, another for Katharine, and have given one to Leonora. I cannot believe that the man who could inculcate maxims of virtue and wisdom so well, could be the sensual and false man that he is represented to have been. Be that as it may, the precious stones he has strung together, and so well set, will always retain their value.

All here unite in affectionate love to you and Charles.

Ever my dearest Fan,

Your affectionate father,

LEONARD HORNER.

To his Daughter.

London, 1*st November*, 1860.

MY DEAREST SUSAN,—It rejoiced me much to hear of your safe arrival in Paris, and that you had found your dear kind Aunt Nancy waiting for you ; and I am most glad that you have found her, your Aunt Fanny and her husband, so well.

I met Sir Edward Ryan two days ago, who made most kind inquiries after you, and spoke in the warmest terms of your book, said he had read it with much interest, that it had come out at the right time, and that he thought it of much importance that it should without delay be translated into *Italian*, and be distributed as widely as possible throughout Italy ; that it would do immense good there.   I suppose that you are in correspondence with the Pulszkys on this subject, but it would be well to urge the speedy accomplishment of this.   I said to Sir Edward that I had found Charles Lyell under apprehensions that our Government is leaning far too strongly in favour of Austria, and his reply was to this effect, " You need be under no apprehension on that score, I suppose you are alluding to Lord John's letter which was quite unauthorized by the Cabinet, and for which he was taken to task by his colleagues."   Sir Edward has the best opportunities of knowing accurately what is passing.

Ever dearest Susan,

Your affectionate father,

LEONARD HORNER.

To his Daughter.

London, *November* 23*rd*, 1860.

MY DEAREST LEONORA,—I often think of the pleasant time I passed in the Behrenstrasse.

I resumed my regular occupation at the Geological Society the day after we got home ; there is much to do there, and I am glad to be able to devote my time to a work that promotes

the cause of science. The work is heavier because of the way
in which we are situated with our secretaries. It is of consequence that they should be men of some reputation as
geologists for the credit of the Society, both at home and
abroad ; and, in Huxley and Warrington Smyth, we attain
that end ; but both, with the best wishes for the prosperity
of the Society, are so much occupied with their professional
duties, that they find it almost impossible to do anything
besides ; Huxley in the Government School of Mines, Smyth
there also, together with some half dozen other official employments. I have thus, in addition to the general superin-
tendence as president, to do the greater part of what belongs
to the secretaries. But as it is a labour of love, it sits very
lightly upon me, even to the extent of a real pleasure. One
of the heaviest duties the president has to perform is to prepare
an address to be delivered at the Annual General Meeting in
February ; and as I am a slow worker and the field to be gone
over is very wide, that preparation occupies me much. We
had an excellent meeting two nights ago, when David Forbes,
brother of the late Edward Forbes, read a paper giving a very
graphic account of the geology of Bolivia, and part of Peru,
where he has been resident three years. Yesterday was the
dinner of the Philosophical Club, and we had a good party.
We had an account of some very interesting researches in
chemistry by Professor Bunsen of Heidelberg, who has discovered a fourth metallic alkali. The Royal Society, as an
acknowledgment of the great services he has rendered for
many years to chemical science, have awarded to him the
Copley Gold Medal, the highest honour the Society has to
bestow. Of the estimation in which this is held abroad you
may judge from the value which Humboldt put upon it. He
received it in 1852, and George can tell you of the box containing it found after his death. Louis Mallet comes back
with a very favourable impression of M. Rouher, the Minister

of Commerce, with whom Cobden has had to carry on the
negotiations about the commercial treaty. He says that no
one could have conducted himself with more perfect fairness
and straightforward honesty, and it has been completed upon
very equitable terms for both countries. He, Louis, is full of
admiration of Cobden. There will be discontents in both
countries for some time among those whose *immediate* gains
the treaty cannot fail to encroach upon, but the principle of
free trade is so sound, that before long the beneficial effects of
the treaty will be substantially felt, and it will render war
between the two countries year after year more improbable.
The visit of the Empress to Scotland is a curious affair, *it is
said*, that she is a very great fanatic, and a devoted worshipper
of the Pope, and that her husband has encouraged the trip in
order not to be bored with her importunities while he is
arranging matters for curtailing the temporal power of his
Holiness. She has been warmly received in all places she
has yet visited.

I am very well, and am careful to bear in mind that I am
on the verge of seventy-six.

My kindest regards to George, and kisses to the three
darlings.

I am ever, my dearest Leonora's

Affectionate father,

LEONARD HORNER.

From Mr. Charles Darwin.

Down, Bromley, Kent, *December 23rd*, 1860.

MY DEAR MR. HORNER,—I must have the pleasure of
thanking you for your extremely kind letter. I am very
much pleased that you approve of my book,* and that you
are going to pay me the extraordinary compliment of reading

* Origin of Species.

it twice. I fear that it is tough reading; but it is beyond my powers to make the subject clearer. Lyell would have done it admirably.

You must enjoy being a gentleman at your ease; and I hear that you have returned with ardour to work at the Geological Society. We hope in the course of the winter to persuade Mrs. Horner and yourself and daughters to pay us a visit.

Ickly did me extraordinary good during the latter part of my stay, and during my first week at home; but I have gone back latterly to my bad ways, and fear I shall never be decently well and strong.

With many thanks for your very kind letter, pray believe me, my dear Mr. Horner,

Yours very sincerely,

CHARLES DARWIN.

P.S.—When any of your party write to Mildenhall I should be much obliged if you would say to Bunbury that I hope he will not forget, whenever he reads my book, his promise to let me know what he thinks about it; for his knowledge is so great and accurate, that every one must value his opinion highly. I shall be quite contented if his belief in the immutability of species is at all staggered.

CHAPTER XIII.

1861.

From Charles Darwin.

Down, Bromley, Kent, *February* 14*th*, 1861.

MY DEAR MR. HORNER,—I must just thank you for your note, but I will take advantage of your kind and considerate offer of discussing the points referred to, till we meet. The latter point seems to me very intricate, and I have often thought it over.

Man does not cause any variations, he only accumulates any which occur ; I do not suppose that God intentionally gave to parent Rock Pigeon a tendency to vary in size of crop, so that man by selecting such variations should make a Pouter, so under nature, I believe variations arise, as we must call them in our ignorance, accidentally or spontaneously, and these are naturally selected or preserved from being beneficial to the successive individual animals in their struggles for life. I know not whether I make myself clear.

Believe me, my dear Mr. Horner,

Yours very sincerely,

CHARLES DARWIN.

———

From Charles Darwin.

Down, Bromley, Kent, *March* 20*th*, 1861.

MY DEAR MR. HORNER,—I am very much obliged for your Address, which has interested me much. I have been particularly glad to see your excellent summary on metamorphism, for I was very ignorant of the recent researches. I thought that I had read up pretty well on antiquity of man, but you bring all the facts so well together in a condensed form, that the case seems much clearer to me. How curious about the Bible ! I declare I had fancied that the date was somehow in the Bible. You are coming out in a new light as a Biblical critic ! I must thank you for your remarks on the origin of species (though I suppose it is almost as incorrect to do so, as to thank a judge for a favourable verdict), what you have said has pleased me extremely.

I am the more pleased as I would rather have been well attacked, than have been handled in the namby-pamby-old-woman style of the cautious Oxford professors.

I most sincerely hope that Mrs. Horner is a little better ; and with my kindest remembrance to all your party, pray believe me, my dear Mr. Horner,

Your sincerely obliged,
CHARLES DARWIN.

Emma sends her very kind remembrances.

———

From the Marquis of Lansdowne.

Lansdowne House, London, *Thursday*.

DEAR HORNER,—I must thank you for having remembered me by sending a copy of your Address to the Geological Society, as delivered last month, and the more so as I can with truth add, having just finished perusing it, that I have never read a statement so clear, and convincing, so much substantial information condensed into so small a compass.

It is quite remarkable to find so many proofs discovering themselves at the same time, and proclaiming the antiquity of man, in a way that it is no longer possible to doubt, and I am glad to be able to indulge that opinion without having my scriptural faith impeached, even in days when Oxford Reviewers are not permitted to escape.

Believe me, yours truly,
LANSDOWNE.

From Dr. Dawes, Dean of Hereford.

Deanery, Hereford, *March* 21*st*, 1861.

MY DEAR MR. HORNER,—Many thanks for a copy of your Address, which I have read with great interest, and it is an additional pleasure its being sent by you.

I like much what you say about Darwin, and his theory, and also about Baden Powell, and I see you venture to quote the essays and reviews notwithstanding the anger of the Bishops and their condemnation of them in Convocation. Most of all I am glad you have expressed yourself as you have done in the few last pages of your Address, on the Chronology of the History of Creation, as given in the marginal notes of the Bible. If you saw as much as I have done of the old routine mode of catechizing children on this subject in our schools, you would feel the more confident in what you say is the origin of the widespread belief in this country, " of the recent origin of man ; " this, I have no doubt, is the fixed belief in a liberal sense as regards the 4004 years, of nine-tenths of the children, if not of all, when they leave our national schools. It is the regular training of our national schoolmasters, and there is no tendency whatever as far as I have seen on the part of Diocesan inspectors or of Privy Council inspectors, to correct it. It is in fact most painful to listen to it.

I hope, however, your observations on this subject will be taken up by the educational and other journals, and that some steps may be taken to correct this erroneous teaching. Perhaps the Bishops when they next meet in Convocation, will class you with the authors of the reviews and essays— don't be too confident that some of them will not.

Mrs. Dawes joins me in kind regards to Mrs. Horner, your daughters, and yourself, and

Believe me, my dear Mr. Horner,

Always sincerely yours,

RICHARD DAWES.

From Mr. Gladstone.

Downing Street, *March 26th*, 1861.

MY DEAR SIR,—On receiving your address I sent a formal note of acknowledgment. I have now read it with great interest, and I wish I were more competent to appreciate the great topics which it opens. I heartily, however, wish well to all who are engaged in the careful collection of facts, and in concluding with caution from them, and feel that if any fault or error arise, it will come not from the observance of these rules, but from failing to observe them.

With many thanks,

I remain, very faithfully yours,

W E. GLADSTONE.

From the Archbishop of Canterbury. [*Summer*]

Lambeth, *April 1st*, 1861.

DEAR SIR,—It is owing to an accident that I have so long delayed my acknowledgment of your kindness in sending me your Anniversary address, full of new (to me) and interesting observations.

You are, I believe, aware that I have always considered the first verse of Genesis as indicating, rather than denying, a *preadamite* world. And if one may venture to speculate on such unknown ground, it would seem more reasonable to believe that any former system of things should contain creations having the qualities of man, than the contrary.

Therefore it would be no surprise to me, if further enquiry should confirm what is at present a new and unexpected discovery, and certainly meanwhile, I shall adhere to Paley's rule, and not suffer what I do know, to be distracted by what I do not know. An axiom, in which I am sure that you will concur.

I remain, dear sir, faithfully yours,

J. B. CANTUER.

From Dr. Fowler.

Salisbury, *April 10th,* 1861.

My Dear Mr. Horner,—I have encroached on my conscience by so long delaying my thanks for the most interesting paper I ever read.

Your observations on Usher's note on the first chapter of Genesis, is, I have no doubt, the cause of the clamour which the over religious public had raised against Geologists for writing up the mountain to be older than Moses.

Egypt and Damascus are the two places in which I should expect more remains of the human race than are likely to be found elsewhere, but when we know that gold was to be found in such abundance, we must not be surprised that the remains of extinct animals, and of our own rude forefathers, should so long have escaped observation. The causes of most phenomena may be found to be nearer to us than we are aware of. Aristotle had the *sagacity to suggest* that objects, *objective Phenomena* may project species to be perceived by us subjectively. But *Hobbes,* one of the most acute minds that has ever appeared, rejected this as little less than a childish supposition, but the photographic process has vindicated the sagacity of Aristotle : a copy of a face distant some feet from a camera obscura, and projected into it, as copies of mutual lovers may be projected into each other's eyes, and thus solve the enigma of love at first sight. The source of terrestrial magnetism appears to me to be nearer to us than the supposed spots of the sun, and that terrestrial heat, a vast source of electric force, may be the source of terrestrial magnetism.

I have sent you four corrections of copies of some of the abstracts of papers which were read for me at the British Association.

Ever, my dear Mr. Horner,

Your obliged,

R. M. Fowler.

From Arthur P. Stanley, Dean of Westminster.

Oxford, *April 29th*, 1861.

My Dear Sir,—I have to thank you for your interesting and instructive address. Every scientific man who writes as you have done, calmly and reverentially on the relations of religion and science, does a substantial service to each ; and it gives me sincere pleasure to feel that you do not agree with the bishops in denouncing all inquiries into these matters, as inconsistent with Christianity, nor with many of your lay brethren, in refusing to clergymen the privilege of writing and speaking the truth.

Believe me to be, yours very faithfully,

Arthur P. Stanley.

To Sir Charles Bunbury.

17, Queen's Road, West, *June 3rd*, 1861.

My Dear Charles,—I am very sorry, as we all are, that we are not to have you and dearest Frances on the 10th, but you give good reasons why you and she should not leave home. I trust that the intended quiet will restore her.

I too am deep in Buckle. As was natural I was anxious to see what he says about Scotland, so I began with that part, and have read as far as the 388th page. I am much interested, and, with few exceptions, go entirely along with him in all his denunciations of the bigotry of my countrymen under the narrow domineering spirit of the clergy, even to the present day. It is frightful to think of the cruelties these ministers were guilty of, for so long a time. I have been very much struck with the defence of himself in the really admirable pages from the middle of 322 to the end of 329. If you read these it will dispose you, I think, as it has done me, to have a more favourable opinion of the author. His industry and patience in reading all sorts of out-of-the-way books in collecting his evidence, is astounding. We shall have the

blasts of many trumpets from the ministers in the north, who
would gladly make an *Auto de fè* of this tremendous assailant.

I am not so much of a Zoologist or of an Ethnologist, as to
feel an interest in de Chaillu's book, and with my Italian
studies, to which I am faithful, Roscoe's Lorenzo, and Buckle,
to say nothing of my Geological *duties*, I have all my time
filled up.

With best love to dearest Frances.

I am ever, my dear Charles,

Affectionately yours,

LEONARD HORNER.

To his Daughter.

Folkestone, *July 1st*, 1861.

MY DEAREST KATHARINE,—Many thanks for yours from
Dundee Harbour. You give a good account of all the
Drunkilbo party. I doubt not that you are all very happy,
and that our dear boys and Rosamond are enjoying themselves
greatly.

You appear to have heard of our expedition to the Preventive
Service Station on Sunday, by mentioning Necker and asking
about him. He is a Genevese, grandson of the great Alpine
traveller and Geologist Sanssure, and of the Minister Necker
of Louis XVI. He was at Edinburgh University early in this
century, and I knew him well. He, from some strange freak
and aversion to all his Swiss kith and kin, settled of all
places in the world, in the Isle of Skye, and I heard of his
going round about dressed in Highland costume and playing
the bag-pipe. The sailor's wife told me she knew him well by
sight. He has lived all that time (now more than thirty
years), at the house of John Cameron in Portree, there being
a Miss Cameron, who makes his supper for him at four in the
morning, his usual time of taking that meal. The report in
Skye is that he is to leave her his fortune, to the truth of

which her matutine services would seem to lend some support. Annie Macdonald, the Sailor's wife, worked in the *Woollen* factory at Portree, the only factory in the Hebrides, which it would have been my duty to have visited, had I continued Inspector for Scotland. I have thus replied to your inquiries about Necker and Portree.

This evening we have been to Saltwood Castle, and then to the camp at Shorncliffe, a most beautiful evening. Barley harvest has begun : we saw the reapers in two fields. To-morrow, George, Nora, Mary, Joanna and I go by train to Dover, thence by the new line to Canterbury and back the same way. We start about one o'clock, and expect to spend three hours in Canterbury.

God bless you all, my very dear Kate,

Your ever loving father,

LEONARD HORNER.

[Mr. Horner on going abroad for his wife's health, sent a regular journal to his daughters at home, much of which is given in the following pages.]

Paris, *September 19th*, 1861.

My dear wife having undergone severe suffering from neuralgia for nearly eight months, expressed so much dread of the risk of a return of this severe complaint, by exposure to such another winter as the last, we determined to seek the milder climate of Italy, and pass the winter at Florence.

We left London the day before yesterday, that is my wife, and daughters Susan and Joanna, and myself, and put up at the Pavilion Hotel, Folkestone. In the evening we found Frank Marcet, his wife and daughter just arrived, on their way to Genoa, *via* Paris. It was a beautiful evening, the nearly full moonlight reflected on a smooth sea, no wind and therefore the prospect of a good passage next day. We had a

smooth passage, but a slow one, for we were two hours and
twenty minutes. After the usual hurry, we got comfortably
settled in a railway carriage, ourselves, Mrs. Clough, who
came with us from London, and the three Marcets filling it;
these carriages giving ample accommodation to eight persons,
and we reached Paris at a quarter to six. A small omnibus
brought us to the Hotel du Congrés, Rue du Colysée, where
my sister, Mrs. Power, had secured rooms for us.

Paris, September 23rd, 1861.

Yesterday was cold and wet, and we did not go out before
half past one, and went to call on Signor Ruffini, the author
of the interesting novel Dr. Antonio, but he was not in Paris.
To see something of Paris we drove through numerous streets,
entering the Boulevard St. Denis by the Port St. Denis,
thence to the column of the Bastile, and came home by the
Rue St. Antoine and Rue de Rivoli. A drive of this sort calls
continually for expressions of admiration and wonder, admiration of the beautiful town, and the display of the shops, with
all the wares so tastefully laid out, and wonder at the vast
crowd of busy people. We hear much of the poverty of the
inhabitants, that they have such enormous rents to pay, and
heavy taxes, but we see nothing that indicates it; on the
contrary, endless evidence of wealth. We passed the evening
in the Rue Ponthieu.

Macon, September 26th, 1861.

The day before yesterday, I walked through the Champs
Elysées, now very appropriately so-called, being laid out so
beautifully, to the residence of the Papal Nuncio in the Rue
de l'Université, to have the Pontifical visa to our passport.
Yesterday we took leave of our kind sisters and Byrne; and
we started at eleven and arrived at Dijon at half past five,
where we dined, and reached this place where we found
excellent rooms provided for us, as I had bespoken them.

The country is pretty all the way, and appears well cultivated. I did not notice any of the small divisions with great varieties of crops which one sees so frequently in Prussia, but a gentleman travelling with us said that the land is much divided nevertheless. The change of soil to the jurassic formation, with frequent bold escarpments, adds greatly to the beauty of the scenery near Dijon, and a splendid sunset so brightly gilded the landscape that we were reminded that we were in the *Côte d'Or*. There were few autumn tints of the foliage, and the vineyards were in bright verdure. We had a gentleman and his wife in the carriage with us, going to Lyons, and from his conversation, I concluded that he is connected with the trade of that town. We of course spoke of the effects of the American War upon it, which he said is very severe, for one-third he said of the manufactures of Lyons goes to the United States. He said that they have nothing to fear from the competition of England under the new treaty.

Turin, *September* 30*th*, 1861.

We yesterday held a consultation about our progress hence, the result of which was, that instead of going by Piacenza, Parma, Bologna, we shall have more pleasure by going to Genoa and there engaging a vetturino to take us to Pisa by Sestri and Spezia. Susan went early to spend the day with the Pulszkys, and Anne and Joanna made a round of the town. I strolled out and went to the Contrada del Po, of which I have a distinct recollection from the impression made upon me when I was here in February, 1817. I walked along the bridge to the right bank of the Po, in the hope of seeing Monte Viso, but the sun shone so brightly on the mountains, that I could hardly trace the range of the Alps. Anne and Joanna were set down at the Pulszky's where we were to dine, and where I joined them at the appointed hour of five. They have a very pleasant house in the outskirts of the town,

looking on the Collina, and the Superga in the distance. We
had a very pleasant dinner, all the family being at table
including the Hungarian young lady, who performs the part
of bonne, governess, and house-keeper, the daughter of a
clergyman, who came to them a short time before they left
England. She was dressed in the costume of her country,
which is very graceful. Gusti and Gabor have also assumed
the usual dress of Hungarian boys. I talked with Madame
Pulszky about Hungary's prospects, and *her* belief is, that if
in six months Austria does not yield, there will be a revolution in the Austrian dominions. They had an evening party
and I was pleased to meet again our Neapolitan friend, the
celebrated Poerio, who is a member of the Italian Parliament,
and at present resides here. He spoke to me in hopeful
terms of a settlement of the southern part of the kingdom, at
no distant period.   I spoke to him about our friend Avisani,
he knew him well, and laments his death, which appears to
have been hastened by his disregard of Medical aid. He too
was a member of the Italian Parliament, but damaged his
usefulness by his impetuosity and impatience of contradiction.
The work on the Popedom, on which he had been engaged
for twenty years, and which occupied him so much in his
daily visits to the reading rooms of the British Museum,
is left I was sorry to hear, in too imperfect a state for
publication.

There was a Mr. Scott with whom I had a great deal of
conversation. He is a manufacturer of firearms at Birmingham, one of the Government contractors. He told me that
the Government provides *the stocks* of the muskets, in order
that they may be assured that the wood is thoroughly
seasoned. He established a work here five years ago for the
manufacture of *stocks* which are sent to Birmingham—the
wood is always *Walnut*, and he comes here because that tree
is to be had in abundance. No tree is fit for their use unless

it is forty years old. He believed that in the five years he has been here, he must have used up not less than 50,000 trees. The supply is beginning to fail, and he is on the look out for some other region to which he must migrate. Our Government in the last two years have bought 600,000 muskets; of these, about 160,000 have been for the Volunteers. I asked how it was that there should be such a demand, as a musket is not a perishable commodity, but so far from this last being the case, he says the average vigorous life of a musket is not more than twelve years. Such is the difficulty of getting all they want, that our Government, besides having forty contractors in England, gets a large supply from Belgium.

This morning we have been to the Picture Gallery, in which there are very many works of the old masters of great merit. Joanna and I went to the Museum of Natural History. There is a fair collection of minerals and other objects, but what interested me most was the Megatherium and Glyptodon from Buenos Ayres, both made complete by very few additions. The shield of the Glyptodon is perfect. Unfortunately both Sismonda and Gastaldi, to whom I had letters from Charles Lyell, as well as one to Sir J. Hudson, are all absent. Glorious weather, I eagerly seek for every bit of shade.

October 2nd.

After a visit to Coutts's correspondent for a supply of the ways and means, I went at the appointed hour, half-past twelve, to call upon Baron Ricasoli, at the *Minestero degli affari esteri*, which is in a wing of the King's Palace. After waiting for a quarter of an hour, when a gentleman came out of the Minister's room, I was shewn in, and the Baron came forward to me from an inner chamber and gave me a most hearty reception. He is a tall man, very thin with a very expressive and benevolent countenance. He asked me if I spoke Italian,

and when I replied in the negative, and adding "I believe you
speak English," he said, "No, and I hope you speak French,"
and on my assenting he said, "Now we shall get on," and he
placed me beside him. He first asked after the Greigs,
particularly Agnes as he called her, and then spoke in very
friendly terms about them. On my saying a few words
about Italy and the interest I take in its regeneration, he
entered upon the subject and continued to speak upon it for
about ten minutes. I wish I could recollect every word he
said, but I can not do more than set down part of the
substance. He said that the revolution is not for Italy alone,
that it is a revolution in the general cause of humanity, that
the object of the King's Government is to establish the
liberties of the country on a sure basis ; that the first thing
upon which all must rest is the promotion of true religion ;
that the governments which have been overturned, interfered
with every one in this respect, to the extent of forcing all to
go to church and of punishing those who neglected to do so,
but that now there is entire freedom, no distinction of creed,
and everyone is left to the dictates of his own conscience.
But this freedom, he added, has not diminished the attendance
on public worship, the churches are visited as before ; there is
no diminution of religious feeling. He only slightly hinted at
the great question of Rome as the great difficulty to be sur-
mounted. There is, he said, a decided improvement in the
state of the south. I asked him if we had any chance of
seeing him in Florence in the winter or spring, and he replied,
"Yes, if I can escape from this place ;" that on the death of
Cavour he was over-persuaded to be his successor, that his
present occupations are contrary to all his cherished tastes
and pursuits, that he now hears far too much of the
méchancetés de l'homme, and that he will rejoice when he
leaves this room never to return to it. I had now been about
twenty minutes with him, and not wishing to encroach farther

on his time, I rose; he then expressed his regret that his occupations made it impossible for him to pay any attentions to us here, but offered to give me letters of introduction to friends in Florence which might be useful to us. He was as good as his word for in the evening he sent to me letters to two of his brothers and to Marchese Feroni, the director of the Gallery. I have been sorry to hear from the Pulzskys that he is on terms with the King very far from comfortable, that there is a *back stair* influence working against him. The King is, however, very popular, and I trust that his error will not endanger the holy cause. His personal bravery is undoubted and has no small influence in the right direction.

Anne and Susan drove out with Mrs. Pulszky, Joanna had gone to the Soperga (it is so written on the direction posts, and not Superga, as I had always heard it called.) After I had finished preparations for the journey of to-day, I set out to endeavour to get a view of the Alps from the *Collina*, so taking a cittadino to the end of the bridge at the termination of the Contrada del Po, I found a path which led me to the Convent of the Capuchines (Capuchi) about three hundred feet above the river and from a terrace in front of the Church had a splendid view of the encircling mountains, Monte Viso towering grandly above the rest; I had to go to the Pulszky's to dinner. We spent a sociable evening with our kind hospitable friends. This visit to Turin has made me think still more highly of that most excellent family than I did before; and that was not little. I was much surprised at the cheapness of the education here. His boys are taught Latin, Greek, Mathematics, History and Italian Composition, for thirty francs a year, and Augusti, who gained the highest certificate at the late examination, has his education free all next year as his reward; his father says that this is the first money he has earned and he shall have it.

We left Turin, much pleased with our visit, by railway at

10 a.m. this morning and arrived at two at Genoa where we are comfortably lodged in the Hôtel Feder.

Genoa, *October 3rd*, 1861.

This morning immediately after breakfast we drove to the Fanale on the right of the harbour and enjoyed the magnificent prospect of the town bay and surrounding hills with the adjoining town of Pieradarena and the shore of the Riviera Ponente. Then we drove to the Palazzo Brignole Sale, where we saw a suite of very magnificent rooms and some very fine pictures, particularly portraits of the ancestors of the present Count Antonio Brignole Sale. We then saw the Palazzo given to Andrea Doria by Charles V., an excellent structure but most to be admired for its situation, gardens and orange plantations, partly loaded with ripe fruit at present. Neither did we fail to see the house he occupied in the interior of the town before he occupied his palace, that house a gift to him from the Council of his grateful fellow Citizens.

Sestri di Levante, *October 4th.*

We started from Genoa this morning at nine o'clock. About four miles on the road at the village of Quarto, we stopped to see the point on the shore where Garibaldi, and his 1000 devoted followers embarked for Marsala on the 5th of May, 1860, and where there is a marble pillar with an inscription commemorating the great event. It is hardly credible that it is only 17 months since that day, which has been followed by such immense and important changes for Italy, and through it, as Ricasoli well observed to me, revolutions in the cause of freedom and human progress. It is impossible to describe in adequate terms the splendour of the scenery, changing its features every few miles. The beautiful Mediterranean was perfectly calm like a sea of azure molten glass. 1 was surprised at the absence of ships, for except a boat near the shore, not a sail was to be seen. I presume that their courses

opposite this part of the coast must be out of sight of land.
We dined at Rapallo, in which all the female inhabitants
seemed to be occupied in making pillar lace. From thence we
came by Chiavari, a handsome town, to this place where we
arrived at half-past six.

Pietra Santa, *October 7th,*
8. A.M.

We started from Sestri with six horses having to pass over a
mountain road, which would require three hours for the
ascent. We soon left the Aloe hedges, and came among
chestnut trees, loaded with their fruit, and these gave way to
bare hills. The road is made with great skill, quite as much as
that over the Mont Cenis. In five hours we reached Borghetto, a
dirty miserable looking place, where we found, however, a very
good luncheon, and saw that the inn had many clean bedrooms, and good beds. Before sunset we reached La Spezzia,
and put up at l'Hotel d'Odessa, a large new inn on the sea
shore, where we were very comfortable. We had the pleasure
to find Mrs. Somerville and her daughters in the house,
having been in the same quarters since June. We passed a
most agreeable evening with them. She is now in her 81st
year, but I found her less changed in her bodily frame than I
expected, as it is ten years since I saw her, and I soon found
out by her conversation, that time had in no degree
diminished her mental vigour. She told me that she was
just sending off some MS. additions to a new edition of her
physical Geography, and that she was engaged on a new
work, the exact nature of which she did not state, but from
the few remarks she made, I found that it was a new line for
her, and that the laws of animal and vegetable life will form a
prominent part in it. She said that in her Editions to her
Physical Geography, she had entered into the recently agitated
question of the antiquity of man, and that she had no doubt that
from the opinion she has expressed on that question and on

the Noachian Deluge, she would be denounced as an infidel by the Orthodox. I was gratified to learn that my late address to the Geological Society had given her information she was not before possessed of, respecting the discoveries of worked flints with the remains of extinct species of animals. I was surprised and sorry to find that she had had no opportunity of knowing of many recent scientific discoveries, as for instance Bunsen's researches on the Spectrum, and the treatise on Biology in the new *Natural History Review*. I yesterday wrote at her desire to Williams and Norgate to send her the last mentioned work, and to Huxley for information on the subjects treated of in the Review, and to Dr. Miller of King's College for some accounts of Bunsen's researches. She does not rise before twelve, but after a cup of coffee at half-past seven, she *works* in bed until that hour. She admitted Mrs. Horner and me to her room this morning, and we saw the nice little old lady in her working dress, with her bed covered with books and manuscripts. Susan and Joanna had much talk with their old friends Martha and Mary, and altogether this interview with friends we so much admire and regard, formed a most agreeable episode in our journey. As we did not intend to go beyond Pietra Santa yesterday, which our vetturino told us would take six hours, we did not leave La Spezzia before eleven, and we reached Pietra Santa soon after five, and put up at the Unione, which has proved a most excellent inn for eating, sleeping, and great civility.

Pisa, October 7th.

Here I am again, after an interval of within a few weeks of forty-five years. It is associated in my mind with the greatest calamity of my life. I lost my father and mother when they were advanced in years, our dear boy Frank we lost when very young, but my brother Frank I lost when he was in the midst of a career of honour and usefulness, at this place, within a gunshot of the spot where I am now writing. Every circumstance connected with his last illness, and our residence here is fresh in my memory. My first object on our arrival was to visit the house in which we lived, and if possible to show my dear wife, Susan and Joanna, the room in which he died. I did not require a guide to the house on the Lung Arno. It was then called Casa Cioni, but is now called Casa Gordon, from having been bought by a lady from Dumfrieshire of that name. It is numbered 727 on the Lung Arno. The good woman who now occupies it, allowed me to see the rooms, and to bring in my fellow-travellers. I had thus the melancholy satisfaction of showing them the room dear Frank used to occupy, and that in which he died.

To-morrow I intend to take them to Leghorn, to show them the spot where his body was laid, and the monument erected by my father to his memory.

Pisa, October 8th, 1861.

We have been to Leghorn, a train took us there in half-an-hour. We found my dear brother's monument almost hid by bushes around it, the inscription on both sides scarcely to be seen. I have given directions for all these to be removed, and a low railing to be placed round the monument. The letters of the inscriptions which are cut in the marble, are scarcely legible, so I said that I should like them to be touched with black paint, but Susan suggested that they should be gilt, and that I have directed to be done. The space between the foot of the monument and the rail will be planted with flowers, and I shall make an arrangement that will secure its being at all times in order. The medallion is in perfect preservation. It is by far the most beautiful monument in the cemetery; it was designed by Sir Henry Englefield. To perform this duty to the memory of my brother has been a source of great comfort to me. We got back to Pisa by half-past two, and visited the Baptistry,

worthy of great admiration. Five years ago it was completely restored, and they have been long engaged at the
same work in the Duomo and Campo Santo, and they are
doing the work admirably well.

From Novi to Massa de Carrara, we travelled through a
country where there is a full display of the rocks of which it
is composed, many most interesting sections. Soon after
leaving Sestri, we lost the Aloe, and we did not again recover
it. The sand from Genoa to Sestri seems to form a concave
mirror which concentrates the sun's rays.

Florence, October 10th.

There appears to be great activity in opening up the
communications of the Kingdom. From the railway (Strada
Ferrata or Ferrovia), that unites Florence and Leghorn, a
branch goes off destined to go to Genoa, it is already open
as far as Viareggio, will soon be at Massa, and will thus
convey the marble of Carrara for shipping at Leghorn. It is
a great object to get it soon to La Spezzia, which is to be a
naval arsenal. They are already at work there, preparing for
docks and ship building. Another railway that is commenced
is to go along the coast to Piombino, and will convey the
produce of the mines of the Monte Catini, and the other
mineral treasures of the Maremma. They are hard at work
in a great tunnel through the limestone of the Apennines
near Pistoia, for the railway that is to unite Florence and
Bologna and Lombardy. The tunnel that is to pierce
Mont Cenis, is in active progress, but I have not yet been able
to collect authentic information about it.

Yesterday morning Professor Savi called upon me, and took
me to the Musuem of Natural History. He gave me a most
interesting lecture of nearly three hours duration on the
geology of Tuscany, illustrated by a new geological map, the
work of himself and of his colleague Menighini, which is
exhibited on a large scale in the National Exposition, also of

the Montè Pisani, of which he gave me a copy. They have clearly proved that the superior beds of the Verrucano belong to the Carboniferous period, for in some works established to work *cinnabar* mines, they found numerous coal plants of which he showed me many excellent specimens, ferns and calamites. To what age the inferior beds belong, which are of great thickness, they cannot say, for no fossils have as yet been discovered, and they are highly metamorphic. This appears to be a country especially calculated for the study and investigation of Metamorphism. The limestone that has been changed into the marble of Carrara is liassic. I asked him if any instance of a junction of the crystalline limestone with an erruptive rock occurs in the Alpi Apuani (the Carrara range), and he replied that nothing of the kind had been met with. I asked to what extent these crystalline limestones extend, and he applied the scale of the map to the region, and it showed twelve Italian miles. It is thus a case of general and not special metamorphism. Of the latter there is a most remarkable case at Campiglia near Piombino, which he described minutely, and laid before me numerous specimens. Here the same lias limestone is converted by contact with a hornblende rock into a marble as highly crystalline as any at Carrara, and he showed me several specimens of the actual junction, admitting of no doubt. The same thing occurs in the neighbouring island of Elba. Not far from Campiglia *granite* has burst through argillaceous rocks of decided *miocene* age. He showed me several specimens of the granite, large grained, with large well-formed crystals of orthoclace felspar, a rock not to be distinguished from the granite of the Mourne Mountains, which it once put me in mind of, as well as of specimens of large grained granite in our typical collection at the Geological Society. If I had been twenty years younger with unimpaired geological legs, I would have gone to Campiglia and Elba to study these grand illus-

trations of metamorphism. Savi gave me a good account of the
proceedings about the Survey ; that a commission was appointed, consisting of most of the eminent geologists of Italy, among
whom he named Gemellaro, Cappelini, and himself as Vice-
President, and that they have made a report to the Government within the last few days only. He does not entertain any
doubt of the Government being favourable, but they can do
nothing until the Chambers meet, and vote the money. The
Museum is very extensive, and in excellent order, the geological
part, the work of Savi and Meneghini, the most prominent
feature.

   We left Pisa this morning between ten and eleven, and in
less than two hours were at the end of our long journey.

                              Florence, *October 12th*, 1861.

   The first thing we had to attend to after our arrival here
two days ago, was to make enquiry about an apartment.
Susan and Joanna went to see the Stewarts, who are paying
a visit to Dr. McDougall the minister of the Scotch Kirk here,
and he recommended a house, where an English family had
lately been. It proved in all respects most desirable. It is
the Casa Fabbiani, in open space opposite the Palazzo Pitti,
the upper rooms looking over the Boboli gardens. We are
assured that it is one of the most healthy situations in
Florence. It is within a short walk of the Uffizii, where
Susan and Joanna will spend a great part of their time, and I
am within five minutes walk of the museum of Natural
History, where I shall pass most of my time.

   We are all very glad of having come to rest, and will not
attempt any sight-seeing, until we have had some days of
repose. We had much rain last night, which has cooled the
air a little, but it is still warm.

                                          October 13th.

   We sat some time with Mrs. Stewart.* Her husband is not

                    *   daughter of Lord Cockburn.

only Minister of the Scotch Church at Leghorn, but has been doing much to promote the cause of the Waldenses. He has persuaded them to remove their college for the education of pastors from Le Tour to Florence, in order that the young men, besides their theological training, should have an opportunity of studying and attending lectures on other subjects. He not only proposed, but effected the purpose for £4000 of a Palazzo large enough to lodge Professors and Students, and to have lecture rooms; and five friends of his in Glasgow sent him the sum immediately on his application ; three of them £1000 each, and two £500 each. They have two Students from Naples, who were priests.

After our visit we walked in the Boboli Gardens.

October 14th.

I did little more to-day than make some calls. The one I found at home was Professor Parlatore. As Joseph Hooker had described him as the Nestor of Tuscan Botanists, I expected to find a venerable old gentleman, whereas an active man without a single black hair bleached, presented himself, met me cordially and proffered every service he could be to me.

I had a walk in the evening along the Via Maggio, crossed the Ponte San Trinitá, passed the Strozzi Palace, more like a fortress than a private dwelling, to the Palazzo Corsi in the Via Tornabuoni to call on the English physician, Dr. Wilson. The streets are placarded with Government pro-clamations in the name of the King, but variously headed " Vittorio Emanuele per la Grazia di Dio e la volonta della Nazione," &c. To one of these titles was added, " Duca di Sardegna e Savoia, Cypro e Gerusalemme," &c. I am much struck with the very fine appearance of the soldiers, of whom there are many of different corps in the streets, generally full grown, erect, and springy, very superior to the French

Infantry. The uniforms look well, and the comfort of the soldier has evidently been well considered. The Bersaglieri, sharp shooters, attract one's notice.

October 15*th.*

Yesterday soon after breakfast your Mamma, Susan and I set out to the Uffizi; that is a great range of public *offices* above which is the celebrated gallery of pictures, sculpture and antiquities. Baron Ricasoli had given me a letter of introduction to the Marchese Ferone, the Director of the Gallery, which I went to deliver, but did not see him, he being engaged at the Esposizione. Leaving your mother and sister in the Gallery, I went to deliver my letters from Baron Ricasoli to his two brothers, Gaetano and Vincenzio. I was directed to the Strada Ricasoli, and the Palazzo Ricasoli, which last I found a very sumptuous house in a short, narrow street, as is the case with most of the Palazzi I have yet seen. Both brothers are in the country, so I left my letters and card. Next I went to the Palazzo Vecchio to deliver my letter to the Marchese Sauli, the Governor of Tuscany. I have as yet seen none of those to whom I have letters of introduction except Professor Parlatore. I returned to the Gallery, I met your Mother and Susan in the Tribuna. I paid due worship to the beautiful Venus de Medici, and admired the other treasures of sculpture and painting in that room, as you may well suppose. In the evening Susan, Joanna, Blanche Clough and I, had a drive to the Cascine. It is the Hyde Park of London, or the Thiergarten of Berlin; more beautiful than the first by the addition of a fine distant view, but inferior greatly in other respects; and superior in all respects to the latter: there were many carriages, among them that of the King's two sons with two gentlemen accompanying them, and no small proportion of street carriages. It was a beautiful evening and we enjoyed it much.

In the evening Dr.* and Mrs. Stewart called and I had much talk with him about the Waldenses, and the progress of Protestantism in Italy. The former are now secure in their worship. There are about twenty thousand in their own valleys, with Churches in Turin, Genoa, Florence and Leghorn, but few of the same creed as yet in other places. Dr. Stewart thinks that the avowed Protestants in Italy, besides the Waldenses, cannot exceed three thousand, and those chiefly among the lower orders, scarcely any among the higher ranks, and the middle ranks fear to lose their trade by the resentment of those above them. I was wrong in saying in a former part of my journal that the Waldenses in Leghorn had gotten a vacant Catholic Church, it was a piece of ground adjoining one. The Roman Catholic faith hangs so loosely upon a large portion of the population of all ranks, that with the progress of freedom and education, as soon as an opportunity offers of dropping it with personal safety, and some more acceptable faith *with external* attractions is presented to them, there will be a material change.

October 16th.

Dr. Wilson called upon us to-day, and as I led him to speak on the political state of the country, he remained fully half-an-hour with us. He is the first I have met with who has taken an unfavourable view of the revolution. He began by stating that great discontent prevails in Umbria; that while under the Papal Government there was no conscription there, and its introduction has caused ten thousand of those liable to serve, "to take to the mountains." That up to this time all has been *couleur de rose*, but the time must soon come when they will have to pay the piper. That the vast expenses of the country since the annexation have been paid not by taxation, but by loans; that the annual expenditure

---

* The Rev. Robert Stewart, Free Kirk Minister at Leghorn, married to the daughter of Lord Cockburn.

has been five times the amount of the regular revenue. That
Tuscany will ere long be in the condition of Piedmont, in
which the taxation is double that of Tuscany, and enormous,
extending almost to everything. That the present price of
the 3 per cent. government paper is 44, having been 72 before
the annexation, (this I can verify by reference to public
documents, and if correct, is the worst thing he said about
the prospects of the country). He said that the *Times* and
other newspapers, in addition to the many inaccuracies of
their special correspondents, have represented Cavour as the
Sir Robert Peel of his country, by the introduction of Free
Trade, whereas he did quite the reverse, he introduced many
protective duties which now exist, as for example on woollens,
glass and earthenware, which amount to a prohibition, for
the advantage of influential native manufacturers, adding
enormously to the expenses of the mass of the population.
When I referred to the great measure of religious freedom
established by the government, in contrast to that of the
Austrian dynasty, he replied that the immediate benefit of that
measure would be felt by so very small a portion of the
population, that it would tell very little, if at all, with the
mass generally. Such were the principal discouraging
circumstances of which he spoke. I hope that I may receive
some alleviations of them from other quarters, as I shall seek
all opportunities of getting good information. As yet we have
only seen the *Nazione*, and *La Gazetta di Popolo*, and
although written in a moderate tone, they are the chief
organs *here* of the new state of things. I must try to find out
if any other paper is tolerated, in which unfavourable views
are expressed. Great events that occur from time to time will
tell us the true state of matters that has preceded them.

October 16th.

The two brothers of Baron Ricasoli on whom I called
yesterday have already returned my visit. They came

separately ; Gaetano, the country gentleman, we were most
taken with.   He is living with his family at their country
seat, but expects to come to Florence for the winter in eight
days, and I hope we may see something of him and his family.
Vincenzio is in command of a regiment here, and is a fine
soldier-like man.

I did no more to-day than take a short stroll to the Piazza
Gran Duca, where there are so many wondrous works of art
brought together.   The outside of the long range of the
Uffizii is adorned with the statues of the great men of Florence
of past times, of which she has such reason to be proud, from
Giotto to Mascagni.

October 17th.

I called at the Museum on Cocchi but he was out, but
I saw Mr. Caruel and Mr. Ancona.   The former is the
assistant of the Professor of Botany, Parlatore.   He was
sitting in a very large and lofty room containing an immense
hortus siccus.   On my expressing my wonder as to its extent,
he said that it is not equal to that at Kew, but *second* to that.

When I came home Signor Passerini called, introduced to
us by Baron Gaetano Ricasoli, as the most learned guide we
could have to all that is most worth seeing in and near
Florence ; Susan had much talk with him about gems, and no
one could have been more obliging than he was in his offer to
assist us.   Before he left, the Marchese Feroni called, the
Director of the Gallery to whom I brought a letter from
Baron Ricasoli.   He sat a long time and talked most agreeably,
especially with Susan on her own subjects.   Although much
occupied with the Esposizione he had made an appointment
to meet her at the Gallery to-morrow morning, to lay open the
collection of gems there for her study, which ordinarily are
only seen through glass cases.   It is impossible that we could
have received more prompt or more effective attention than
we have done from those to whom Baron Ricasoli has

introduced us. Baron Gaetano has sent us an order to admit
us to the Boboli Gardens any day in the week; this will be
a great comfort to your mother, for they are not three
minutes' walk from our house. This evening we have received
an invitation from the Marquis Sauli, the Governor of
Tuscany, to dine with him (we four) next Sunday. He
occupies apartments in the Palazzo Vecchio.

October 19th.

Professor Parlatore called and sat some time with me
conversing agreeably. The Botanical collection has been
formed entirely by himself, excepting the splendid addition of
specimens and books given by Webb. He was appointed
Professor twenty years ago, and met with every encouragement
from the Grand Duke, who built the rooms in which the
collection is contained. Among other things we talked about
the language here, which he says has been much corrupted of
late years by the adoption of French words italianised. I
told him what I had heard from Mr. Caruel's friend, that
there is a district near Pistoia where the language is spoken
with the greatest purity by the humblest classes. He said
that it is perfectly true; it is a hill country around the
village of Marcello, near Lizzarro, where he often goes to
botanise, and he finds the guides who accompany him, speak a
language every word of which he could put down as the
purest Italian both in structure and pronunciation. He is a
native of Palermo, and I was sorry to hear him say that
things are not going on well there.

October 20th.

We dined at the Marchese Sauli's yesterday. He, as
Governor of Tuscany, has apartments in the Palazzo Vecchio,
and they are very handsome.

It was a small party, the only lady being a Marchese
Franzoni, whom he told me is his cousin. Your mother had

the Marchese on one side and an agreeable person on the other, and they both talked French to her. After dinner I had much talk with the Marchese Sauli in English, which he speaks with ease. He impressed me with a high opinion of his understanding, and I am told that he is a man of great accomplishments and is a very learned philologist. He talked very sensibly on geological questions, and told me he had read much of Lyell's and Murchison's works, especially referring to the "Principles," which he greatly prefers to the Manual. He told me, what I was sorry to hear, that he is going to leave Florence, for the office of Governor of Tuscany has been abolished. We talked a little upon the affairs of the country, a subject I did not think it right to do more than lightly touch upon, considering his position, and I was glad to hear him speak confidently of things going on well, even at Rome, where there are men of considerable influence who sympathize with their views. We came away much pleased with our host, with the lady (who is going to call upon us), and with the sight and the associations of the place, thinking of the men who had occupied the Palazzo, and the tale which the great staircase, five hundred years old, could tell of those who had trodden the steps.

From Mrs. Somerville.

Spezia, *October 28th,* 1861.

My Dear Mr. Horner,—I ought long ago to have told you how truly I am obliged to your kindness for recommending and procuring for me the numbers of the *Natural History Review;* they contain some of the subjects on which I was writing at the time of their arrival, so you may imagine how much I was pleased. I shall certainly become an annual subscriber. Then soon after, I received Bunsen's and Kirchoff's German paper, as well as that of the Royal Institution, which have quite astonished me; a new field of research is opened of the most wonderful character, leading to results of which no one can see the end, and which probably no one of the present generation can hope to see, though no doubt many are already making observations, and I hope Faraday is among the number.

I am happy to hear that Mrs. Horner has not suffered from the cool air of autumn after the hot summer, for Florence is much colder than Spezia; here the weather is quite glorious, not a cloud in the sky, and though cool in the mornings and evenings, the sun is so powerful that I sat out in the shade the whole morning yesterday, and shall probably do the same to-day. This hotel has been crowded with birds of passage, chiefly English, on their way to Rome, which they say is to be unusually full. Lady Annabella Noel, however, is not to be one; I heard from my son yesterday that she is to spend the winter with her friend at Basle, but of that I daresay your daughters are aware, as I know they correspond. After the English Expositions, that at Florence must appear insignificant, but you have abundant interest and occupation in the Galleries, but they will be too cold for Mrs. Horner in winter. I nearly forgot to ask if you had heard of the new theory of Mr. Faye, the French Astronomer; he entirely denies the existence of an ætherical medium, and attributes the aneteratives of Encke's comet, and the tails of all comets, to a repulsive force in the incandescent surface of the sun, and he has proved by experiment that all luminous surfaces emit a repulsive force. If this singular theory be true, what is to become of the undulatory theory of light!

Martha and Mary are out drawing from nature, with the Marchesa Doria, or I know they would join with me in love to you and yours.

Your sincere friend,
MARY SOMERVILLE.

Journal.

Marchese Carlo Torrigiani is a very agreeable and sensible man, who takes a warm interest in all that relates to the moral condition of the people. He sat three-quarters of an hour with us and talked most agreeably. We expect to learn a good deal by his assistance.

On Saturday we met, by appointment, Professor Parlatore in the *Museo d'Istoria Naturale*. We went first into his own particular department, the great collection, the hortus-siccus made by himself; next he showed us his collection—a very extensive one—of specimens of all the vegetable subtances of which use is made, and models in wax of their structure. This is similar to Sir W. Hooker's Museum at Kew, and has been formed by himself. On this, as on many other occasions, I lamented my ignorance of botany. What an idle, careless fellow I was in my youth! what precious years and opportunities I missed! I hope that my grandsons will take warning by me, and profit by every opportunity offered to them, and especially to cultivate a taste for Natural History in its chief branches, an inexhaustible source of interest and happiness in every situation in which they may be. We suspended our progress in the Museum to another day, as standing so long had fatigued your mother and me, and Mr. Parlatore took us to the tribune Galileo. This is a kind of temple erected about sixteen years ago in honour of that great philosopher by the late Grand Duke, who, bad as he was, as the ruler of his people, had a considerable taste for science, and was by no means uninstructed in different branches, they say; and I remember hearing Babbage give a favourable report of a conversation he had with him. A beautiful statue, by a living artist, of Galileo, forms the chief feature, and in glass cases are various instruments that belonged to him, among them the telescope he made, and by which he discovered Jupiter's satellites. There are also in cases round the room various pieces of apparatus by which the celebrated experiments were made by the Academia del Cimento, as, for instance, the great lens, by which they showed the combustibility of the diamond. My impression is that it was Smithson Tennant, who, by collecting the produce of the combustion, first proved the diamond to be pure carbon.

Of the many great men whom Tuscany has given birth to, Galileo is, to my mind, the most illustrious, and I have been eager to see all the places and things here that relate to him. Parlatore has lent me the "Life of Galileo," written by Viviani, one of Galileo's disciples, which I am now reading, published 1654, thirteen years after Galileo's death, and there I find distinct contradictions to Arago's assertions, who denies him the invention of the telescope, giving it to the accidental discovery of the Middleburg spectacle-maker, and that too in the face of Bailly, who, in his " History of Astronomy," with which Arago could not fail to be intimately acquainted, says, " Si le veritable inventeur est celui qui cherche avec connaissance de cause, et qui de principe en principe parvient au bout qu'il se propose, Galiléo est *l'inventeur du telescope*." The celebrated story of Galileo's deductions when a youth, from observing the swinging of the lamp in the Duomo of Pisa, which is distinctly recorded by Viviani, Arago treats as possibly *imaginaire*. We have not yet been to Arectri, where he lived after his return from his persecution at Rome, and where he died after five years of total blindness.

Yesterday we went to Careggi, about two and a half miles distant, not far from Fiesole ; the villa of the Medici, where old Cosimo died in 1464, and Lorenzo the Magnificent in 1492. It is now the property of Mr. Sloane, who has repaired and added to it, but the building is as it was originally, the internal hall the same, and the great staircase, and even some

of the rooms. The view from the upper story, where there is
an external gallery all round, is most beautiful.

To-day it has been pouring, but we went to the church of
San Lorenzo, and have seen in the Sacristy the celebrated
monuments of Giuliano de Medici, and Lorenzo, Duke of
Urbino, by Michael Angelo, and beautiful and wonderful
they are. The Medicean Chapel too is most beautiful
in its marble and pietra-dura decorations.

November 7th.

The day before yesterday the Marchese Sauli called here,
and as I was alone at home I had him all to myself, and he
sat a full hour talking on all subjects most agreeably. He
speaks English easily, so that I got on better than I should
have done in French ; as to Italian, I am yet far from that,
although I find that I get on, thanks to diligent reading, and
not often giving way to indolence in using the dictionary. I
read with far greater ease the old writers, such as Machiavelli,
than modern ones ; in the history of the reigns of the
Austrian Grand Dukes down to the last by Lobi, he uses
numerous words which I do not even find in my dictionary.
To get familiar with ordinary conversation I am again
resorting to Goldoni's Comedies. In reading the "Life of
Galileo" by Viviani, written two hundred years ago, I had not
occasion to look into the dictionary more than about half a
dozen times in a closely printed 4to page. Marchese Sauli
leaves very soon for Turin, to attend his duties in Parliament,
which meets on the 20th. Their proceedings will excite much
interest, for we shall learn by the speeches on both sides
what progress is making in the consolidation of the kingdom,
and what measures are being adopted towards that end.
Marchese Sauli laid much stress on the importance of pushing
on the new railways as fast as possible, as the most effective
way of making the Italians known to each other, which

hitherto has been impossible from the absence of the means
of locomotion, and besides, they will immensely increase the
productive industry of the people generally. The valuable
publication of Passaglia has recently been most essentially
assisted by the publication of a letter addressed to him by the
Pope some years ago, upon the occasion of a publication of
his on the subject of the Immaculate Conception, in
favour (God forgive him) of that most stupendous piece
of priestly superstition, which was, of course, quite after
Pio Nono's own heart, and in that letter he addresses
him in the warmest terms of praise for his most
learned *orthodoxy*. The publication of Cardinal Andrea,
giving an account of his resignation of the office of President
of the Congregation of the Index in consequence of the
unwarrantable interference with him as a judge by Antonelli,
and the Pope, cannot fail too to have a strong effect upon the
bishops and higher clergy. It is they who are the opponents
of the present state of things; the humbler clergy are to
a great extent in favour of it. If that be the case, *their*
opinion will tell most in the mass of the people. The state of
the schools is lamentable; that must be one of the great
objects of the government. Marchese Carlo Torrigiani gave
us yesterday a most melancholy account of the small number
of children of the labouring class who attend any school, and
of the wretched payment of the schoolmasters, not exceeding
twenty pounds a year. They are paid by the Communes;
no aid was given by the government. This was the case
under the Grand Duke. The present government have not
had time to organize any new system, and their present
claims are even more urgent for all the money they can
venture to raise. It seems to be a wonderful contrast to the
former state of Florence, for in a work I was reading yesterday, it is stated that in the fourteenth century, ten thousand
children in Florence were attending school.

The government have erected in the Esposizione, two modern statues of eminent Tuscans, one with reference to the political unity and independence of Italy, the other with reference to its commercial advancement. The former of these is Francesco Burlamachi, on whose pedestal is written, *Francesco Burlamachi, Da Lucca, Primo Martire, Della Unita e Indipendenza d'Italia, Tre secoli dopo la morte*, 1861. He was the son of a distinguished Florentine, and his uncle was a contemporary and a member of the same convent of San Marco with Savonarola, whose life he wrote, and a copy of which we have obtained. Francesco, while Gonfaloniere of Lucca in 1546, raised the standard of revolt against the Grand Duke Cosmo I. He appears to have been the first of a race of Protestants who settled at Geneva, of whom an account was published some years ago by M. Eynard. Among others, he mentions our friend De la Rive as descended from a Burlamachi. The other statue is in honour of *Bandini*. On his pedestal is written, *Sallastio Bandini, Nostro, Con i massimi veri, della Scienza Economica, Traviti da lui prima di Adam Smith, beneficio il Gener Umano.* This claim of priority to Adam Smith's "Wealth of Nations" excited my attention, and I asked Marchese Torrigiani on what it is founded and to what part of Smith's doctrines it referred. He told me that the claim rested, he supposed, on a book which he lent me, and which I have read. It is a memorial by Bandini to the reigning sovereign of Tuscany, pointing out the miserable state of the Maremma, in which he lived, being Archdeacon of Sienna. It is a sad picture of the consequences of ignorant interference and oppressive taxation by governments, for there was scarcely a thing which was not taxed, they could scarcely move without the leave of an employé, and were prohibited from exporting their produce. Bandini points out most clearly the evils, and his suggestions for their removal, and for a

the horrors committed upon and by the Moriscos.   His account
of Don John of Austria is very striking, particularly of his
early life; but the humanity for which he gives him credit on
many occasions, seems entirely to have left him when he took
the command in the Low Countries, according to what Motley
has told us of his deeds there.   Mr. Prescott gives a more
favourable view of Phillip's character than I ever supposed
could have been given of so cold-blooded a monster as Motley
makes him.   I must read again the first two volumes of his
reign, but these do not come down to the time when his
cruelties in the Netherlands first began.   If I were younger I
would set about learning Spanish to enable me to read
Mendoza's "Guerra di Granada," and some of the other
authors he quotes.   You must shew me what Mr. Ticknor has
said of Mendoza.

Have you heard whether Mr. Prescott has left his materials
for the continuation of his history of Philip in such a state
that they can be published.   It is sad that the most interesting, at least the most exciting part of his reign, the affairs
of the Netherlands, were not all to come from his hand.

Ever, my dearest Mary,

Your affectionate father,

LEONARD HORNER.

To his Daughter.

17, Queen's Road West, *June 16th,* 1859.

MY DEAREST FRANCES,—I wish I could see your roses, and
these with the tranquillity of the white room, and some opportunity of reading, would be a great contrast to our life here.
I scarcely ever get an hour's quiet reading; I must say however, that I do spend a good deal of time, all I can spare, at
the Geological Society, in the herculean task I have undertaken, to get our museum out of its discreditable disorder.
The task would be easier if I could find some one to help me,

same spot of a Spiritual and Temporal sovereign must give rise to numerous difficulties; if the lay government does not include some of the eminent men of Rome, there will be much discontent; if it does, they will be encumbered with men ignorant of public affairs. He does not think that the publications of Passaglia of the Cardinal Andrea will have much effect in altering the conduct of the bishops and higher clergy. On whatever side we look, immense difficulties are to be overcome in bringing into anything like a harmonious kingdom of Italy the various elements of which it must be composed, and for a long time the wisdom of her ablest statesmen will be severely tested. But if she is let alone by her neighbours, the ends in view are so manifestly for the advantage of the people, that the power of those comparatively few who lose by the change, will daily lessen. I believe, too, that if they were attacked by Austria (and no other power is to be feared), there would be a simultaneous rising and successful resistance.

November 10*th.*

The wind has changed from the north to the south, and has brought a milder temperature, the thermometer in the morning being sixty degrees. We have, however, had a good deal of rain. As to-morrow is Martinmas day, we must hope that we shall have *le petit été du Saint Martin*. I fear that I may have appeared very insensible to sorrow in having scarcely alluded in this journal to the sad episode in the history of our life here, now of a month's duration, the illness, which there is every prospect now of shortly proving fatal, of poor Mr. Clough. But I know that in the correspondence of your mother and sisters it would be much dwelt upon. It has been a great source of satisfaction that they have been able to be of so much use to his poor wife; Susan and Joanna have been indefatigable in her service.

November 14*th.*

Yesterday, at an early hour, a messenger came to tell us
that poor Arthur Clough had died in the night.   Susan and
Joanna went soon after to be with his afflicted widow, and
they will give all the particulars of this sad event.

*   *   *   *   *   *   *   *

We had a visit from the Marchesa Franzoni; she gave us a
good account of the pains taken in Tuscany in the education
of daughters.   We long to know something of female society,
she and Signora Giarre, who gives lessons to your sisters in
Italian, are literally the only two Italian ladies we have yet
come to know.

November 18*th.*

On Saturday evening we went to the Teatro Nuovo to
witness one of the exhibitions peculiar to the country, an
Improvisatrice.   The performer was a Madamigella Giannina
Melli, a Neapolitan, who, Marchesa Franzoni informed us, has
a very high reputation.   It certainly was an extraordinary
performance, even with my very slight power of following her,
for although I am getting on in ordinary conversation, I am
still far from following what is spoken either in the pulpit or
on the stage.   Her first proceeding was to read slips of paper,
on which were written subjects sent to her, upon which she
might declaim.   She read each aloud, and then crumpling each
up separately, she threw them into a glass jar.   I think there
could not be less than seventy or eighty of these slips.   She
then took the jar to the boxes on each side of the stage and
asked those in them to take out some of the papers at random
from the jar.   About a dozen were taken out; she returned
with them to her table and read each aloud.   There were two
descriptions of her performances.   In the one she read a
paper, and sitting down with a pen in her hand she invited
the audience to pronounce single words, having no connection
one with the another.   After she had written down about a

dozen, she immediately rose with the paper on which she had written them in her hand, and without a moment's meditation she gave out with great animation, lines ending each with one of the words, and in the order in which they were written down. The other kind of declaiming was this; she took up one of the subjects, and after reading it aloud to the audience, she walked about the stage in contemplation, sometimes as long as ten minutes, and then poured out a full stream. There were three examples of this, and each could not have consisted of less than a hundred lines. They were delivered with animation, and much use of the arms, but always gracefully. It was certainly a most surprising instance of this singular power, and she was enthusiastically applauded by the audience. Most of the subjects had reference to the independence of Italy, and the name of Garibaldi occurred at least a dozen times. The last subject she took was "Garibaldi and Cavour," a difficult one, but judging from the frequent bursts of applause, she must have been successful. There must have been Codini present, to whom these parts could not have been very agreeable, and *one* of them ventured to express his dissent; had there been more, their voices would have been speedily drowned by counter cheers. That they must be insignificant in numbers can admit of no doubt when the Improvatrice ventured on such a subject in a public theatre.

In the evening the Marchese Torrigiani and Professor Parlatore paid us long visits, and were particularly agreeable.

November 19th.

Yesterday we had a deluge of rain, but as the Natural History Museum is near at hand, and as Parlatore had made an appointment to meet us there to show it to us, Susan and Joanna and I went. We advised Mamma not to expose herself to the damp. The Professor first brought out from his hortus siccus some plants which are figured on the Etruscan Vases, about which Susan had a talk with him last evening.

We then walked through the rooms, of which there are forty,
some very spacious.   In every department there are considerably large collections, all well arranged and exhibited,
and set off with every advantage of handsome glazed cases.
There are most ample opportunities to study, and many
beautiful specimens.   The animals and birds admirably stuffed
and set up.   The last part we saw was the celebrated collection
of wax models, most wonderfully executed.   I should like
very much to go through a course of anatomy with such
illustrations.   A great part was made more than a hundred
years ago, and all are in perfect preservation.

We have a beautiful day, but cold, men going about the
streets with crossed mantles, and the thermometer early in the
morning has been as low as thirty-seven degrees.   Notwithstanding this, I bought in the street a nosegay of sweet
smelling roses, heliotrope and jasmine for threepence.

November 25th.

I this morning paid a visit to the Marchese Gino Capponi,
in his handsome Palazzo in the Via Sebastiano.   I concluded
that I should find him in some handsome apartment on the
first floor, but, on the contrary, I had to mount to the third
story which (by the reckoning as I came down) proved to be
one hundred and nine steps up.   I found him in a comfortable
ordinary apartment.   He was very agreeable, and I was glad
to hear from him, and no doubt he had it from the best
authority, that there is a much greater unanimity in
Parliament in support of the Government than was expected.

December 1st.

The book I am occupied with is Villari's " Life of Savonarola," which is very interesting and admirably well-written.
It is quite a recent publication.   It is not only the biography
of that extraordinary man, but a history of the time in which
he lived and acted so conspicuous a part.   He exposes so
many errors in preceding biographies, that I should not be

surprised to find some mistakes in the article on Savonarola by Sir James Stephen, notwithstanding all his erudition. To-day we went to the great Carthusian monastery, the Certosa, on the road to Rome, and part of the way there were hedges of roses in full bloom. To our great disappointment we were told that no woman could enter the precincts occupied by the holy men. I was admitted, and shown over the whole by a lay attendant. It is kept in the highest order, as if it had been built and furnished last year, and it was most curious and interesting to go over, seeing that such things could be at this time of day. There are thirteen monks and twelve Conservi, all in the white dress of their order; the monks were at Vespers in their gorgeous chapel with closed doors, but I heard their chant; the twelve Conservi in rows of six on each side of the richly ornamented vestibule, bearded and hooded, and bowing and crossing themselves as they heard the sound of certain parts of the chant from the Sanctuary. I saw all that is described by Murray, not excepting the suite of rooms that were occupied by Pius VI. in his flight from Rome in 1798. He remained here eight months. The monks never leave the place, and, except one day in the week, observe entire silence. When such things are still permitted to exist, it is very evident that the superstitions of the Roman Catholic Church have a long life before them.

December 5th.

Since I wrote last I have seen two more convents, one the celebrated San Marco, Dominican, the other, the Franciscan at Fiesole. I was desirous of seeing the former, as the residence for many years of Savonarola, during which he exercised such immense influence over the people and destinies of Florence. It has besides a large number of the frescoes of Fra Angelico during his twenty years' residence as a monk, after leaving the Dominican convent of Fiesole, the place usually attached to his name. Many of them are wonderful

productions, not for the time only, but for any time, from the
expressions in the countenances of the multitude of figures.
I was in the cell so long occupied by the great reformer.
There is nothing very remarkable in the chapel. I was
disappointed to find the resting places of Pico di Mirandola
and Poliziano marked only by slabs of marble. I was alone,
because the greater part of the convent is not shown to
women. I could not see the library, which is very extensive,
as the Frato who keeps the key was absent.

Yesterday we went to Fiesole, where Susan and Joanna had
not been. We could not have had a finer day for the view, for
the atmosphere was so perfectly clear, that the buildings on
the distant hills appeared within a short distance, and the
sun was unclouded, but there was a sharp north-east wind,
which made wraps of all kinds in the open carriage very
necessary. The views were perfectly beautiful, the contrasts
of the grey olive trees, the dark green cypresses, and the
autumnal, or rather wintry tints, of the deciduous trees, were
most striking. We had to ascend a steep part of the hill on
foot to the Franciscan convent on the summit, which we
visited, the ladies as far only as their feet were permitted to
tread, I over the whole, which is very extensive. Here there
are forty-six monks, one of whom, a man of seventy, showed us
round. I think that we may well apply to them, and to all
such fraternities, what Sir William Hamilton said of the
ministers who left their livings to join the Free Kirk movement, that they are martyrs by mistake.

December 19*th.*

One day last week we went to the Palazzo Medici, long
known by the name of Palazzo Riccardi, that family having
bought it in the early part of the sixteenth century, adding
much to its extent. We had a letter to the Librarian of the
celebrated Riccardi Library, and to the Secretary of the
Academia della Crusca, which still holds its meetings in this

Palace. The Librarian was absent, but an attendant showed us some of the remarkable books, several beautifully illuminated classics and breviaries, but what interested us most was Savonarola's *own* thick duodecimo Bible, with the margins in many parts covered with his notes, in writing so small, that even strong eyes require a magnifying glass.

December 21*st.*

Your dear mother's seventy-fifth birthday, and most gratefully do I say that I have not seen her looking better, or in better spirits, for a long time. She only this day declared her readiness to mount to the tower of the Palazzo Vecchio to see the dungeon of Savonarola before his execution in the Piazza below. It has been a beautiful day, but a sharp north-east wind; when I rose at half-past seven there was a bright, rosy sun-rise, and we had a blue sky all day. We have been to the Villa Mozzi, near the summit of Fiesole Hill, a favourite spot of Lorenzo de Medici, beautifully described by Hallam. The modern fittings-up are in very good taste, and there are a few very good pictures; we were most charmed with Lorenzo's terrace, and the splendid view from it.

December 27*th.*

Among the many institutions of this place for the promotion of literature, science and art, there is one called the "Istituto di Studii superiori pratici e di Perfezionamento," in which lectures are delivered by Professors, who are said to be more or less distinguished in their several departments. Having read with much interest the "Life of Savonarola," by Professor Pasquale Villari, and seeing that he was to lecture on the Filosofia della Storia, and the lectures being public, Susan and I went about ten days ago. We found a very capital and appropriately furnished room near the Lorenzo Chapel, and a considerable audience. Being desirous of making Dr. Villari's acquaintance I got a letter of introduction to him from Marchese C. Torrigiani, and called at his house. He was at home and I passed a very agreeable hour with him. He spoke English with great fluency, although he told me he had never been in England. He spoke of his intimacy with Lacaita, he himself being a Neapolitan, and of John Stuart Mill as a particular friend. He showed me two folio MS. volumes of unedited works of Savonarola, chiefly commentaries on the Old Testament, especially Genesis, Isaiah, Ezekiel, and the prophets Amos and Haggai. He was in the habit of covering the margins of his private Bible with notes, in so small a hand that it is very difficult to read. This very Bible I saw in the Riccardi Library. So small is the writing, that a portion of one of his MS. which Mr. Villari showed me, about the size of half a playing card, occupies three folio pages in the volume that I saw, the hand writing and distance of the lines of the transcript very little more than those in this letter. Paper must have been very dear in his time.

Mr. Villari drank tea with us last night, coming at eight and staying till near eleven, during the whole of the time his conversation being most entertaining and instructive. I hope that we may see him often.

All that we were told of the bitterly cold winds of Florence has been fulfilled to the letter within the last week, so your mother and I keep very much at home. I must not forget to add that we have much bright sunshine, and that for ten days we have had no fog.

CHAPTER XIV.

1862—1864.

To his Daughter.

Florence, *January* 19*th*, 1862.

MY DEAREST MARY,—I thank you for your affectionate congratulations on my seventy-seventh birthday, and you could not have sent me a more acceptable present than your own portrait. Mamma made me a present of four lithographs in a case, of Power's busts of Washington, and Franklin, the Fisher Boy, and the Diana. He is certainly a great artist, and a pleasing, simple mannered man. He has in his study a great assemblage of the busts of remarkable Americans, and with few exceptions they are not Apollos; Mr. Everett's is very good. I am much obliged to Charles for his interesting letter, the geological part of which I will soon reply to. Your account of your dinner at the Duke of Argyll's we read with much interest, especially what you tell us of your conversation with the Duchess about our poor Queen. Her strength of mind is most wonderful, and if anything could increase the admiration and love of her subjects, her conduct in this the most trying occasion, the greatest she could go through, would do so. I have long thought her the best monarch that ever sat on England's throne, and I think that no one would now dispute it. We are most happy to learn that Lady Clark has rallied, did you ever hear whether Sir James got a letter I wrote to him on hearing of the death of the Prince?

Yesterday we had a sharp tramontana, and to-day we have a fall of snow.

Best love to Charles. God bless you, my darling.

Your affectionate father,

LEONARD HORNER.

To his Daughter.

Florence, 21*st January*, 1862.

MY DEAREST KATHARINE,—You have heard that I have been much interested by reading Villari's history of the " Life and Times of Savonarola"; so much so, that I thought it very desirable that there should be an English translation of it, and I was led to think of undertaking it myself. The Marchese Torrigiani introduced me to Villari. I told him what I was inclined to do, provided there was no other translation in English, already existing or in progress. He said that he knew of neither, and that he should be most happy if I would undertake it; adding that he should be glad to give me any assistance in his power. With the encouragement he gave me, I set to work. Many thanks for your kind congratulations on my birthday, if I am to see another or more, I trust that it will always be with a large number of my children and grandchildren around me.

Your affectionate Father,

LEONARD HORNER.

Journal.

Florence, 4*th February*, 1862.

The Marchesa Torre Arsa, the wife of the Prefect, called to-day, and sat some time with us. She is one of the most agreeable persons whose acquaintance we have made. I asked her about the procession of Sunday. She said that the threatening of it had caused her husband some anxiety for three days. It was caused by an assertion of Antonelli that the people of Italy are not against the temporal power of the

Pope, that it is only the people of Turin who are so. This document was most industriously circulated by two violent Codini papers here, who added to it some violent abuse of the King, the Government, and some of the leading men here. We have heard that the numbers could not have been less than thirty thousand. Not a single outrage was committed· The Prefect sent for Dolphi, a *baker*, who has great influence with the working classes, is most humane and cautious, a strong liberal, but averse to all violence, and a powerful speaker, and he assured the Prefect that no outrage would be committed. Not a pane of glass was broken in the houses of the most violent of the Codini.

To his Daughter.

Florence, 11th February, 1862.

MY DEAREST MARY,—It seems long since I wrote to you. We had seven days of mild agreeable weather, your mother and I went to the Boboli Gardens and the birds were singing as if spring had arrived; but the very next morning when we awoke, the roof of the opposite house was covered with snow, the distant hills high and low, have white cloths on; and there must have been a good deal on the mountains for the mails were retarded considerably. My friend, John Russell, at Edinburgh, is gone, at 82. I had a great regard for him. We have been intimate now for nearly forty years, ever since the foundation of the Edinburgh Academy; the success of which was, in a considerable degree, owing to his zeal and judicious exertions. Except Pillans, I do not know that I have now an old friend left in Edinburgh, I believe I am the last remaining of the first body of Directors of the Academy. But it is the course of nature that the ripe fruit should fall, but how many we have lately seen long before, arriving at that state. . . . .

I had a letter from Sir James Clark this morning, with a

somewhat better account of his poor wife.   It is written from
Osborne, and tells a good deal of the poor Queen.   She has
become, he says, thin and weak, but he is not anxious about her
health, and the Princess Alice is much in the same state.
The arrival of the Princess Royal, expected on Saturday, will
be a great comfort to them.   For a long time the death of her
husband will appear more and more afflicting.

We are all well, your mother is wonderfully so ; I am going
on steadily with my work, but it moves slowly.

Your affectionate father,

LEONARD HORNER.

————

To Sir Charles Lyell.

Florence, 14*th February*, 1862.

MY DEAR CHARLES,—We had a visit from Gino Capponi
yesterday.   He is by far the ablest man I have had the good
fortune to meet here.   Notwithstanding his long total blindness, he is up to all that is going on in the political world,
and has most sound and liberal opinions, accompanied by a
sagacious moderation.   We had a long conversation with him
about the great demonstration last Sunday week.   I thought
it a good thing, and a good sign, but he soon satisfied me it
was to be deplored for the sake of the good cause of Italian
independence : he said, and he has the best opportunities of
being rightly informed, that the manifestation was *got up* by
the democratic party connected with that society of Genoa to
which Garibaldi once belonged, but from which he some
time ago withdrew, as he held their proceedings to be most
dangerous to Italy's regeneration.   He told us that the
manifestation here, and at some other places, has drawn
forth an excellent letter from Ricasoli to the Prefect, and an
excellent article in the *Opinione* of Turin, the organ of the
government, that we should find both in the *Monitore* of that

day. I have read both with so much interest, that I send the paper to you.

I have read your letter to Murray about the Prince with much pleasure, if it is not to be published in any form, it will be much to be regretted.*

Yours affectionately,
LEONARD HORNER.

To his Daughter.

Florence, 24th February, 1862.

MY DEAREST MARY,—You have been very kind in writing to me very often since we left home, and thus have done all you could to comfort me in my separation from you, which all the charms of Florence do not compensate for, neither as regards you or any one I left behind, but I have a full compensation in the restoration of the health of your dear mother. I rejoice to hear that you have seen some amendment in Mrs. Twisleton, her spirits about her own country must be kept up by such letters as you have been receiving. I am perfectly lost in reading what the newspapers contain, I can form no definite idea how matters stand, and I know not what to believe, I can see nothing before me, but the certainty of many years of embarrassment and unhappiness in that once happy and prosperous land. Everything there is on so vast a scale that their misfortunes seem to bear the same proportion as other things. . . . . I met with a very intelligent man at Mr. Sloane's last week, a native of Quebec, and he told me much about Canada that was very satisfactory. He said that the French population are most firm in their attachment to England, and so are all those of English origin, but that it is very far from being the case with the Irish. What a great object it might be for an

° It is published in the Memoirs of Sir Charles Lyell.

English statesman to probe to the very root the causes of discontent among so large a portion of the people of Ireland, and boldly apply the remedies.   There was nothing your poor uncle had more at heart, and had he lived, I do not doubt that he would have effected much: One of the last things he did was the amendment of the law about Grand Juries, a measure that gave satisfaction to *all* parties, a sure accompaniment of any measure of improvement there.

I will immediately read Dean Milman's article on Savonarola. The pamphlet of Villari I was sure would interest you.   His views are to me in many instances very original and are suggestive of much thought on the philosophy of Italian and European civilization.   I should like to have the judgment of such a mind as that of Hallam or Macaulay upon it, but where are minds of such a cast now to be found! that many exist I doubt not, but they do not stand prominently out.

This has been a charming day.   I have been walking in Torrigiani's garden, and found the rose bushes putting forth their leaves, and a camelia shrub in the open air with many flowers fully out, and covered with bursting buds.

It is now about time for me to prepare for bed, nearly ten o'clock, for we keep early hours, and the days have now lengthened so much that I can be up by seven.

Good-night, my dearest Mary ; mother and sisters quite well.

Your affectionate father,
LEONARD HORNER.

Journal.

13*th March*, 1862.

To-day we have seen one of the most interesting things in Florence, the Galleria Buonarotti; the house in which Michael Angelo lived when he visited Florence in his old age, and where are collected many most interesting memorials of him, preserved with great care, set off to the best advantage ; a gift to the City a few years ago, by a Buonarotti, that those relics might be kept together and never dispersed. There is a wonderful group of sculpture, when he was only seventeen years of age, before he made the celebrated mask of the Faun, and there is the last letter he wrote when he was ninety-eight years old. But I did not thoroughly enjoy it, the rooms and the stone floors were so cold. It is most strange that a town which has not less than six months of temperature requiring mantles and *scaldini*, should have built their houses as if they were at the equator, the greater proportion of their rooms without a fireplace, and not one room with a wooden floor.

31st March, 1862.

Lord Harrowby, who has been here a fortnight, and whom we have seen several times, has introduced us to a most singular person whom he has known for many years; a Mr. Kirkup, who has lived here more than twenty years. Susan and I visited him this morning in his home at the south end of the Ponte Vecchio, which belonged to the Knights Templar. We found a very old man with a white beard, occupying some rooms through which we passed, filled with all sorts of antiquarian objects, including vast collections of books, sundry instruments, three parrots, and birds flying through the room ; himself very like an astrologer, and Lord Harrowby tells me he is deeply versed in spirit-rapping. We found him very agreeable, spirited and liberal, and we shall often visit him for he has much that is curious to show. He is intimate with our friend Villari. He has some curious things relating to Savonarola which I shall profit by. Lord Harrowby has promised to send for my inspection, when we return, a most curious picture of the Friar, by Fra Bartolomeo, that he has at Sandon Hall.

12th April.

Yesterday Joanna and I went to Leghorn, my chief object

was to see the repairs of my brother's monument, which I
had ordered six months before. I found all quite to my mind,
the shrubs that hid it have been removed, an iron railing
placed round it, the marble cleaned, and the letters of the
inscription gilded. I trust that those who are to come after
me, will not allow it to be neglected. No more burials are
now in this cemetery, and the Committee of the English
Church are having it put in complete order. We got home
after sunset, we had a very bright day, and much of the way
the scenery is pretty, the plain being bounded on both sides
by hills.

14th April.

We paid a visit this day to the antiquary, Mr. Kirkup. He
seems the personification of a *virtuoso* in the best sense of the
term, and he seems to have a vast number of rare and curious
editions. He showed us the first edition of the "Gerusalemme
Liberata," that was printed while Tasso was in prison; an
authentic plaster bust of Macchiavelli, and many other
curious things. He it was who discovered, and got cleared off
a thick coat of many repeated whitewashings, on the walls of
the Chapel of the Bargello, the fresco with a portrait of Dante
as a young man. He saved from destruction the doorway of
the house in which Dante was born, close by the little church
of San Martino, where he was married, and which has been
the place of assembly of the twelve *Buonuomini* for the relief of
the *poveri vergognosi* from their institution in the early part of
the fifteenth century, by San Antonino, to the present day, a
blessed establishment, deserving of imitation in every country.

To his Daughter.

Folrence, 18*th April,* 1862.

MY DEAREST MARY,—I go on with my translation of
Savonarola very steadily; I had yesterday an interesting
walk with Villari; he took me to see an original portrait of

the Friar by his brother monk in St. Mark's, Fra Bartolomeo, in the possession of a Signor Rubieri, painted a very short time before the Friar's death. The history of the hands through which it has passed is a most curious one. We next went to the convent of St. Mark's, and there I was shewn some things not commonly shown, the mantle, tunic, and hair shirt he wore on the day of his execution, and some other relics. I was in a room with about a dozen of the monks, who seemed to be reposing between one service and another; and in appearance and dress, Villari said, I might fancy myself transported back to the very place and company of Savonarola. I have collected several books relating to him, and a most curious print of his execution in the Piazza del Signoria.

Your affectionate father, LEONARD HORNER.

To his Daughter.

Florence, *29th April,* 1862.

MY DEAREST FRANCES,—We were three days ago at Mario's Villa, formerly Villa Salviati, that possessed by the Archbishop Salviati, who was hanged out of a window of the Bargello as one of the chiefs in the Pazzi conspiracy. We were in the room where it was [hatched. It is by far the most beautiful and most interesting villa we have yet seen. This evening we are going to Fiesole. Your dear mother is well in all respects, except that unpleasant double vision which Zanetti, whom we have consulted, says is a stomach affection. Your sisters are well. My side gives me pain when I walk, but Zanetti thinks it must be rheumatism.

Barton, doubtless, is now in great beauty, and will be growing more and more so for a long time now. I trust that I shall see it about the same time we were with you last year. My love to Charles.   Your affectionate father.

LEONARD HORNER.

[Sad days followed; a telegram came on the 2nd of May announcing the death of a precious grandchild in Berlin, from gastric fever. Soon after Mrs. Horner caught cold when driving in the Cascine, and great weakness ensued. In spite of sanguine hopes which the physician gave, who advised her to be taken out drives (which she did till the 15th of May), she sank on the 22nd, to the overwhelming sorrow of her husband and daughters. She was buried in the cemetery at Florence, and a marble monument was placed over her grave. After a most blessed union of fifty-six years of uninterrupted devotion, the blow was most heavy, but he tried to look back on his happy years. As he wrote to one of his children : " I will dwell on the recollection of them over again with her, deeply grateful for their long duration, to her, as well as to myself, but it is a severe trial to go through—a dissolution of such a union." On leaving her remains in Florence he writes : " It is a spot soothing to look upon. Florence will be now connected with us by an indissoluble link. Although I know that her blessed spirit is in that state of bliss which a loving God has assigned to the pure, it will not be possible for me to disconnect her in my memory with that spot on earth. We must accustom ourselves to look with gratefulness on her long, happy life, on her calm removal from it, and feed on the belief that she is in a blessed state of 'existence, and the hope that before long we may meet, never to part."

In the middle of June (with Sir Charles and Lady Lyell, who had come to him immediately on hearing the fatal news) they all left Florence to return home. Among the many letters he received, the following gives a true picture of his beloved wife. Having sent a copy of the inscription on the grave to his old friend Mrs. Burge* (Margaret Alison) she writes:

Daughter of the Rev. Archibald Alison.

Woodville, Colinton, Edinburgh, 11th June, 1862.

Many thanks for letting me have that precious copy of the
inscription on one who was so pure and holy an example of
all that is best and holiest in life. Her gentle and loving
spirit seemed to me always to influence all around her, and to
throw the mantle of her own blessed nature upon those who
belonged to her circle. What must the loss be to those
nearest and dearest to her. I can only assure you, dear
friend, that we have sorrowed with you with our whole hearts,
and have felt for all that you are feeling. Yet many mercies
and alleviations have been granted you in these hours of trial,
and I am thankful to hear you are able to feel them even
now.

The reflection of her happy life must be a never-failing
balm to your heart, surrounded and guarded as she was by
love all her life; and to the last, save in the loss of that little
one who led the way to Heaven, she was blessed and happy
in every relation. She expressed this happiness to me when
I last saw her, so strongly and so thankfully, that it made a
deep impression on my mind, and now recurs to me with
double force.

May God bless and keep you and all so dear to you; and
bring you in safety back to your own country. I believe she
is still near you, and may this belief also be yours, and your
children's, so that you may feel the loving spirit is only a little
way behind the veil; and that before long you shall be again
united for ever   With our kindest love and truest sympathy,

Ever affectionately yours,

MARGARET BURGE.

Mr. Horner on his return to London took a house in
Montagu Square, where, after spending the summer with his
family at Eastbourne, he settled in November. He spent
Christmas at Barton.]

To his Daughter.

Barton, *21st December,* 1862.

MY DEAREST MARY,—It was a great comfort to me on this sorrowful morning,* to receive your affectionate letter. It was mercifully ordered that when this day last year I wished your angelic mother many happy returns of it, the idea even that it might be the last never entered my thoughts, so well was she, so improved in health, so youthful. That my dear absent children should to-day each have expressed their mournful thoughts and affectionate sympathy in mine, has been a source of consolation to me, for I gladly welcome every occasion when we can commune upon her virtues, her affections, her every quality that endeared her to us so intensely. We must cling to the hope, to the belief which all analogy concurrently warrants, and nothing of any weight discourages, that her blessed spirit, that is, *her own self,* on quitting the bodily frame in which she dwelt, passed into a state of blessedness where we shall meet again. To think of her, to speak of her, to cherish the remembrance of whatever will bring her before us, will be the best consolation for her loss that is left to us, and a great one it will ever be.

I am glad to hear that Charles' cold is better. I brought Huxley's lectures to the working classes, with me. I have read the first, it is most excellent, and to me most instructive. They will be an excellent preparation for another reading of Darwin's work, which I am contemplating. I have looked at Darwin's new work on the orchids, but have read no more than the introduction. The concluding sentence of page 2, places two classes of readers in contrast, whereas I think they may be united by the second observing " Did not the Creator make the secondary laws? Is not every trifling detail of structure a part of the original design, if it be of constant and not accidental occurrence ? "

° His wife's birthday.

I am glad to hear that Charles' long expected work,* has advanced towards completion by another chapter.

God bless you, my dearest Mary,

Your affectionate father,

LEONARD HORNER.

To his Daughter.

Barton, 24*th December*, 1862.

MY DEAREST KATHARINE,—I wish you a happy Christmas, which it will be, if your husband and children are well ; a *merry* one it cannot be, with such a weight of sorrow as you have upon your heart, when you think of the departure of that blessed spirit which was always the life of our circle, and which must be so much in your thoughts. I have Italy almost constantly in my thoughts, and in one shape or other, it will, I believe, be what my mind will henceforth most dwell upon. My occupation with Savonarola has, happily for my comfort, led me to a certain familiarity with the country, and the mournful, calamitous event is a link that will bind me to it for the rest of my life. I am reading Macchiavelli's " History of Florence ; " and besides that, a book I have found here, Trollope's visit last spring to a part of the country seldom visited, the ancient Umbria, a part of the late States of the Church, which he entered from Arezzo to Città Castello, thence to Gubbio, Perugia, Assisi, Foligno, Camerino, Macerata, Ancona and Rimini ; very curious and interesting, but terrible scenes of the cruelties practised, and the feuds of the numerous chiefs who ruled over the poor people in the middle ages.

My love to Harry and the dear children.

God bless you, darling,

Your affectionate father,

LEONARD HORNER.

° " The Antiquity of Man."

To his Daughter.

Barton, 31*st December*, 1862.

MY DEAREST KATHARINE,—May you, Harry, and all the
children see many happy New Year's days; and may that
which is to commence to-morrow, end without any affliction at
all approaching to that which the expiring year brought upon
us, or sorrow of any kind.   To-morrow the recollection of what
your adored mother was last New Year's day at the Prefect's
ball ; so well, so handsome, so full of enjoyment, will make
me feel severely the sad contrast.   We must cherish the
well-grounded belief, that the blessed soul is in full possession
of the joys for which her virtues and well-spent earthly life
were so sure a preparation.   This has been a beautiful day,
and with the exception of one day, we have had enjoyable
weather since Christmas day.   I hope you read Gladstone's
speech at Chester in the *Daily News* of yesterday.   It is quite
admirable.   Frances and I have been to Bury.   She is looking
well, but she worries herself with her duties.   Charles is very
well, and does not allow his duties, which he discharges not
less effectively, to disturb his tranquillity.

Dearest Katharine,

Your affectionate father,

LEONARD HORNER.

To his Daughter.

London, 17*th February*, 1863.

MY DEAREST LEONORA,—I trust both your darlings will get
through the trying time of spring weather without any illness.
I hope you have got by this time the copy of my " Savonarola."
What with the " Antiquity of Man," " Savonarola," and Mr.
Byrne's Memoirs, you will have enough to occupy you for some
time.   I have finished the first reading of Charles's book, and
have begun the second reading, for there is much in it that

requires careful study. In a few days I shall have another work that will be very interesting, Professor Huxley's "Evidences as to Man's place in Nature," I expect to see a very near relationship to the gorilla being made out. On this subject see a remarkable passage in Hallam's introduction to the "Literature of Europe," Vol. IV., Chap. II., Section 44. There is a very nice popular volume which you should get: Huxley's six lectures to working men.

Charles Darwin was in town for a week, looking, as he usually does, quite robust, but so liable to illness that he could not venture to come to us either in the evening, or to breakfast. But he is full of work, and read a paper on the plant *Linum* at the Linnæan Society last week, which has excited much attention.

Colenso has brought forward to Englishmen, the learned and philosophical essays of De Wette, and many others of the German theologians, and added much of his own that is interesting. The Bishops and Bibliolaters may rave, but until they answer him successfully step by step, they will not diminish the effect which his work is producing, all for the advancement of *true* religion.

Ever, my dearest Nora,

Your affectionate father,

LEONARD HORNER.

To his Daughter.

60, Montagu Square, 11th *June*, 1863.

MY DEAREST FRANCES,—Your dear affectionate sympathizing letter this morning is just like yourself. Yesterday* I spent with your dear sisters, Susan and Joanna, tranquilly. We went in a carriage to the Crystal Palace, the day was favour-

---

° His wedding day.

able and there were comparatively few people, and we walked in the garden and looked at the beautiful objects inside, and spoke of your darling mother and of my last visit to the place on the same day of the year, two years ago with her.   I felt very sure that reading her letters would be an occupation of the warmest interest to you.   They have been a source of much happiness to me, for I have lived my life of happiness with her over again, and I shall often have recourse to them for the same enjoyment.   In my daily looking upon her dear pictures, I think with deep gratitude upon the great sum of happiness in a long life for which I am indebted to her, not only as regards her own sweet companionship and her excellent sense, as a wise counsellor, but as the mother of my dear children, all reared by her with the best principles and affections.   For all these unspeakable blessings, I render humble thanks to the Father of all Goodness.

I look forward with much pleasure to paying a long visit to you and your dear kind husband.

I am ever, your affectionate father,

LEONARD HORNER.

_____

To his Daughter.

Edinburgh, 11*th July*, 1863.

MY DEAREST KATHARINE,—We are delighted at the prospect of having you, Leonard and Rosamond here next week, all our plans while you are here will have reference to you, and we promise ourselves great pleasure in going over the town and neighbourhood.   You must stay over the 22nd, which is the day of the Academy Exhibition, which will be a pretty sight for you, Leonard and Rosamond, and I will lose no time in obtaining tickets.—I had written thus far when a packet was delivered to me containing a request by the Directors that I should preside at the Exhibition day and deliver the prizes.

I shall make out a plan for the week. I am making notes for my geological expeditions with my dear Leonard.

Your affectionate father,

LEONARD HORNER.

To his Daughter.

Edinburgh, 24*th July*, 1863.

MY DEAREST FRANCES,—Our visit to Edinburgh has been very pleasant, we have seen many kind friends, and looked with much interest on old familiar places. We have paid two visits to the Miss Gibson Craigs at their charming villa of Hermiston. I have had much gratification at the Academy, finding it in so excellent a condition, and meeting with so warm a reception from my friends there. The Directors invited me to the dinner they gave, and first the Lord Advocate said some very kind things about me, and then Lord Neaves (one of the Judges) proposed my health in terms that my most ardent friends could not fail to be satisfied with, and which were received in a very kind way indeed by the company. Katharine, Leonard, and Rosamond having been here has been a great increase to our pleasure. I shewed Leonard some of the more remarkable of the geological features of Arthur's Seat.

My love to dear Charles.

Your affectionate father,

LEONARD HORNER.

To his Daughter.

Edinburgh, 14*th August*, 1863.

MY DEAREST LEONORA,—I am looking forward with much pleasure to the prospect of soon seeing you with dear Annie and Dora, and am very sorry that George's strong sense of official duties should deprive him and us of the additional pleasure of his bringing you to England. You seem to have been most fortunate in choosing Coblentz for a part of your holiday. I believe that the Mendelssohns are now at Horch-heim, and if so, give both my kindest regards. It is now very nearly thirty years since I left Bonn as a residence, and I always look back with much satisfaction to the time we spent and the lasting friendships we formed there.

Our visit to your and my native land, has been most agreeable in many ways; we have received the most hearty welcome from all our old friends who are still living, and from all those who are still of an age that entitles them to look forward for many years. I have had great satisfaction in my visits to the Academy, and have even been of some use, for I am in close communication with the Directors about some important improvements in our system of teaching. The School of Arts, too, I have found very prosperous, and I have this day been engaged about a scheme that will add to its usefulness. Our visits to Drumkilbo and Shielhill were most agreeable. Next Tuesday we hope to reach Barton.

Ever, my dearest Nora,

Your affectionate father,

LEONARD HORNER.

To his Daughter.

Barton Hall, *4th September,* 1863.

MY DEAREST MARY,—I was glad to learn that the visit to Newcastle had been so agreeable. I have seen the reports in the *Times* and *Daily News*, and was much pleased with Sir William Armstrong's discourse, and from your account, besides being a man of no ordinary information, he has a very generous spirit. I have always thought Crawford's views against the unity of the human race very conclusive, and his reasonings for the great antiquity of man, from the growth of languages, very sound. We are all well here. In the last week there have been many visitors, and we have had fourteen and sixteen to dinner, hence a duration of the repast, and more indistinct talk than is agreeable to me. I lead a very regular life. I have a breakfast in the literal sense, by a cup of cocoa, and generally an hour and a half's pleasant reading before the family repast at ten. I go to my room immediately afterwards and work and read until luncheon-time, before which I take a stroll in the Arboretum or pleasure-ground for half-an-hour. Then drive in an open carriage for an hour and a half and get about three hours' reading before dinner at half-past seven. I am generally *in bed* by half-past ten. Can I do better? "Yes," you will say, "when you can have a quiet dinner,"—agreed, perfectly.

I have been doing Academy work, Egyptian work, Geological Society's work. There is nothing *wrong* in the Academy, but room for improvement in order to adapt the system of instruction to the altered demands since it was settled forty years ago, and I had a most satisfactory talk on the subject with Lord Neaves, who takes a very active interest in the school, and with Mr. Comrie Thomson, the present intelligent and energetic Secretary.

I am summoned to luncheon, which is early as we are going to Bury. Love to Charles.

My dearest Mary's affectionate father,

LEONARD HORNER.

———

To his Sister, Mrs. Byrne.

60, Montagu Square, *12th December,* 1863.

MY DEAREST FANNY,—I have to thank you for your kind letters of the 5th and 10th inst., and am glad that you and dear Nancy are well and not suffering from the inconstant weather so liable to be felt by us old people. I am very

happy to hear of the additional testimonies you are receiving
to the value of your dear husband's memoirs; the letters of
Rutherford Clark are good and creditable to him. I was
particularly glad to hear that Lady Bell had heard Lord
Monteagle speak in strong terms of commendation of the
work, for he was very capable of forming a sound estimate of
its value. I received a short time ago, a copy of the report
of the Committee on the financial affairs of Ireland,* addressed
to me in a handwriting I did not know. I immediately read
the report attentively, and with great interest, and no small
surprise. It is drawn up with evidently great care and in most
temperate language, and does present a most flagrant case of
wrong to Ireland. It cannot fail to make a strong impression.
I could only guess at a *probable* friend as the sender, and
guessed Chief Baron Pigot, and wrote to thank him for the
gift. In doing so I said that I hoped a copy would be sent to
every member of both Houses of Parliament, and that a
summary of it might be sent to every newspaper in the United
Kingdom.

I sent my copy to Louis Mallet, and it is to go from him
to Edward Romilly. It is a wrong far more likely to be
acknowledged and soon remedied, than that of the Episcopal
Church, and I hope that it will be brought forward in both
Houses by men of authority and calm judgment, not connected
personally with Ireland. It was very kind of you to write to
Mr. Dillon to send me a copy, rightly judging that it would,
as a document of facts, greatly interest me, and I beg you
to express to Mr. Dillon how much obliged I am to him.

I have just finished a second and more attentive reading of
Renan's " Vie de Jesus," and it has left upon me a much more
favourable impression than my hasty first perusal did. It is
a valuable addition to the works of German *lay* authors on
the same subject, who have treated it as a question of history,

By John Dillon, Esq.

and free from the foregone conclusions which even the most honest *clerical* writers cannot throw aside. There is no want of reverence for the character of Jesus, on the contrary, there is more sentimental imaginings of his character than is quite consistent with the gravity of a work professedly historical, and indeed there is an amount of sentimentalism throughout the work that places the author not in the first rank of French writers of history. I am not at all surprised at the fury of the priesthood, for their *craft* has not been more powerfully assailed for a long time. But pure Christianity, that which Christ himself established, applicable to all nations, and which if fully acted up to, would give peace and good will to the whole human race, will be more firmly rooted in the minds of all who will exercise the reason God has given them, and obey the dictates of that conscience which He implanted in us *as a part of our nature*, to be our sure and safe guide.

We are all well, the last accounts from Berlin were equally good. In a letter I have had from Pertz* he expresses himself as much gratified by the great honour conferred upon him by the Institute. A greater testimony to the value of his literary achievements could not have been given. It is especially gratifying to those who know the great modesty of the man and his laborious devotion to literature *for its own sake*.

To his Daughter.

London, 15*th December*, 1863.

MY DEAR FRANCES,—It is a long time since a letter has passed between us. I believe you have been passing a quiet Darby and Joan life the most of the time since you returned to Barton, but that you are again plunged into the vortex of Christmas hospitalities. May they prove as enjoyable as

* Chevalier Pertz, author of the "Life of Baron Stein," and Editor of the *Monumenta Germanica*, who married his daughter.

those autumnal ones we had the pleasure to partake of. The
presence of Edward just fresh from Transylvania and Asia
Minor must be a great event, and it must be no ordinary
gratification to your guests to listen to such subjects as he
can discourse of, by so good a talker. The recent wonderful
discoveries at Athens are of intense interest, especially as they
may lead to something even more important than the
architectural treasures.

For ourselves I have nothing to relate, for your sisters
and I go on very quietly, and, except dining at Harley Street
on Sunday alone, we have not dined out since we were at Sir
Edward Ryan's a fortnight ago. In this way I pass evening
after evening in the agreeable society of my books. Having
read Milman's "History of the Jews" to the end of the reign of
Herod the Great, I took up Renan's " Vie de Jesus," which I
read attentively a second time, having only skimmed it before.
He gives in some respects a similar history to what I read
before in German works, with the addition of a more detailed
geography of the country in which Jesus lived, and He in
some instances is wanting in due reverence. But there is a
continual sentimentalism very unsuited to the dignity of
history. I shall resume the far more agreeable style of
Milman.

I was much pleased with Lord Russell's excellent practical
reasons for not agreeing to go to the Congress, I am very
sorry for the Poles, and for Denmark, but I cannot give up my
time from pleasant occupations to read the endless newspaper
articles about them.

Interested as I am in the defeat of the Slave-owners of the
Confederate States, I am contented with a short summary of
the progress of the Federal; and as a political barometer, I
look at the price of the Confederate Loan in the money
market, and am glad to see that sure index in the minds of
the money speculators, of a happy cause. I read every word

in the newspapers about Italy, and rejoice over the sure signs of progress and of the ultimate triumph of that righteous cause.

My best love to Charles, and kind regards to the Eastern traveller.

Ever, my dearest Frances,
Your affectionate father,
LEONARD HORNER.

To his Daughter.

Montague Square, 21st December, 1863.

MY DEAREST KATHARINE,—I am very sorry I was out when you were here to-day.   I could not fulfil the proposal of going to Twickenham to talk on indifferent subjects when my heart was sad.   Joanna took me to Westminster Abbey, where I had not been for many years.   Your uncle's statue pleased me more than ever, and I think it is one of the best, if not the best in the Abbey.

Thank you for the flowers to be placed before your adored mother's likeness, as they have been on this day* which for so long a period was to us all, the happiest of the year, when we saw the bright face of that beloved one, the centre of our happy circle.   Now we must look back with thankfulness for these blessed days, while we cannot, and would not if we could, forget that she is departed from us.   Susan did not fail to call my attention to the heliotrope in your bouquet, as a favourite flower of your darling mother.

Love to Harry and your children.

Your affectionate father,
LEONARD HORNER.

* His wife's birthday

To his Daughter.

Montague Square, 27*th December*, 1863.

MY DEAREST FRANCES,—The weather has become much colder, and I am very sensible to its influence, but I am well clothed and muffled up when I go out. People at my advanced age are often carried off by sudden colds and attacks of bronchitis, and it is necessary to be careful.

We had an admirable sermon from our minister, Mr. Foster,* yesterday, on the Christian duty of being of a cheerful disposition in gratitude to God for our numberless blessings. He rivetted my attention by the soundness of his arguments, and the eloquence with which he enforced them.   There was much beautiful music ; there is a fine organ, and it is admirably played, and several well-trained singers.   We had the Pastoral Symphony of Handel, and the service closed by a hymn sung to the music of " Sound the loud timbrel," &c., a composition quite worthy of being Handel's.   May you and dear Charles see many New Year's Days in health and happiness.

Your affectionate father,

LEONARD HORNER.

————

1864.

[The cold weather brought on cold and cough, and though during the month of January he was able to go out occasionally, all February he became weaker, and it was a sad time for all his family.   His mind was quite clear to the last, and perfectly aware that he could not recover ; all his children were with him, those at a distance having arrived early in February, and his loving thought for each of them was always present, his unselfish life remained strong to the last, always thinking of others and trying to spare them.

* At the Free Christian Church.

On the morning of Saturday, the 5th March, on being offered some liquid in a spoon, he said "I will try," and his beautiful spirit passed away.

He was buried in Woking Cemetery, in the spot he had selected as a family burying place, his six daughters and four sons-in-law, and two young grandsons followed him to the grave. A beautiful marble monument was raised over the vault in which he was buried. It was sent from Italy, and it bears the following inscription :—

LEONARD HORNER, F.R.S.

BORN IN EDINBURGH, 17th *January*, 1785.

DIED IN LONDON, 5th *March*, 1864.

Love of truth and love of his fellow creatures, were his characteristics.

He was during fifty years an active and energetic member of the Geological Society of London, of which he was twice President, and made valuable contributions to the Science of Geology.

He was the Founder in 1822, of the Edinburgh School of Arts, for the instruction of mechanics.

He was one of the two Founders of the Edinburgh Academy.

As Government Inspector of Factories for twenty-five years, he laboured zealously for the welfare of the working classes.

Ever a devoted son, brother, husband, and father, he passed away, loving and beloved, with his faculties bright to the end.

"It is Heaven upon Earth, to have a man's mind move in Charity, rest in Providence, and turn upon the poles of Truth." LORD BACON.

"Let us not be weary in well doing, for in due season we shall reap if we faint not. As we have opportunity let us do good unto all men."

Galations vi., 9, 10.]

LEONARD HORNER.

(By George Ticknor, Esq.)

From the Boston (United States) "Daily Advertiser,"
29th March, 1864.

The last mails from England have brought us news that will be sad to many Americans; the death, in London, of Mr. Leonard Horner. He was full of years—seventy-nine we believe. Certainly he was full of honours, and of the love and respect of friends in many parts of the United States and of Europe. Among us, not a few knew and valued him as the biographer of his brother, Francis Horner, a statesman whose early death is still to be counted among the misfortunes of his country, and whose life, re-published here in 1853, has served to form and strengthen the principles of many an aspiring young jurist in the United States, as it has in England, from its first appearance there. Others on our side of the Atlantic have known Mr. Horner as a naturalist, who at one period was president of the Geological Society, and who contributed, at different times, valuable papers to its transactions. And still others have known him personally as the father of Lady Lyell, to whom and her eminent husband, so many Americans became attached during their visits to the United States, and who were always proud to present to their distinguished father the friends from abroad who visited them in London. The late Mr. Prescott, the historian, was one of them; and how he enjoyed a visit to Mr. Horner and his family is plain from one of his charming letters to Mrs. Prescott, in which he gives an account of a day or two he spent with them at their little villa on the Thames, near Hampton Court; and of a gay *picnic* on

Box Hill, of which, or other parties like it, more than one American friend has still the most pleasant recollections.

But in England, although Mr. Horner's scientific character was always recognised and honoured, he probably attracted a more general regard by his marked qualities and powers as a wise, good man, efficient for practical service in society and the State. In this character he had much to do with the organisation and first movement of the London University, and he was for many years one of the leading and most active "inspectors" under the Factory Act, doing much personally to ameliorate and raise the condition of the working classes in the great manufacturing towns in the West of England, and in other parts of the kingdom. Indeed, whenever anything was to be done for the improvement of society Mr. Horner was always ready to contribute his part and money; and he always did it wisely and well.

This happy activity continued to the last. His faculties never failed; his affections never grew cold. The death of Mrs. Horner—a most attractive, cultivated, and faithful lady —which occurred about two years ago at Florence, where the family had been spending some time for her health, was indeed a shock from which Mr. Horner never recovered. But he did not weakly yield to it; on the contrary, to occupy his thoughts during the dark season of anxiety and sorrow, he translated, with happy fidelity, Villari's remarkable life of the marvellous Savonarola, to which he added notes, and published it a few months ago. Work, therefore, consoled him to the end; work always for the advancement of knowledge, and for the good of others; and, when the inevitable hour came that comes to all, it found him surrounded by loving children, and thankful for the blessings of a long and useful life, during which he had enjoyed, in a degree granted to few, the affection of the many who knew him well and the respect of all who had in any way come within the sphere of his influence.—G.T.

[Extracts from some of the many letters of sympathy which poured in to the family:

From the Rev. J. J. Tayler.

" I shall ever cherish with a kind of reverential tenderness the memory of your excellent father.   It is now, I suppose, some fifteen or twenty years since I first became acquainted with him.   I found that on many points my views coincided with his, and that he was·actively engaged in a very important, but not always a very popular work in Lancashire.   I could give him perhaps more sympathy than he met with in every quarter.   His work.is now done.   It must be a satisfaction to his children to recollect, that according to his conception of what it was, he discharged it with high honour, unremitting zeal, and a fearless single-mindedness.   He is gone where purity of intention and rectitude of endeavour will meet their fitting reward."

From Lady Bell.

" Dear children of beloved parents, let me join with you in thankfulness.   The long life of *happiness* ended two years ago: your dear father was still alive to the love and gratifications from his family, but he never recovered the loss of his first love.

" How I comprehended and admired his *bravery* and his acceptance of the mercies around him! but who can wish better for those we love than an easy release from the struggle. This has been granted to your father, and his children may be thankful that they have been his comfort.   My old and best friends are departing so quickly, that I have no time to mourn. Indeed, since the rending asunder of my own self, this world has been to me merely a passage."

From Sir Louis Mallet.

" I know too well the strange and painful solemnity of these separations from those with whom our whole lives have been identified and associated, and the thronging of tender associations which crowd upon the heart and brain at such moments. A little later you will feel as we did, what a blessing it is where there is so little either in the life or death to regret, when the recollections that remain are those of unmixed love, veneration, and gratitude, and when the hope of reunion is bound up with all our highest aspirations for the life beyond the veil."

From Mrs. Adams.

" I feel, when I think of his long, pure and good life, his sweet, tranquil death, all his and your happiness for all these years, as I did when my own darling father died at eighty, that it was all a blessing, we could only feel gratitude."

From Miss Johnson.

" If the better world *now* exists, you are sure that he is thither gone. His life was so innocent, so useful, and so happy, that we can look back upon it with firm confidence."

From Professor Sedgwick.

" After some delay the mourning card, announcing Mr. Horner's death, reached me at Norwich. I received the news with much surprise and sorrow for I had not heard of Mr. Horner's illness. He was the last remaining friend of my early geological life. I experienced from him many acts of good-will and kindness. I associate him in memory with Buckland, Conybeare, Greenough, Fitton, John Taylor, and

some others.   All are gone, and I am remaining like a dried
log that shews the high-water mark of byegone days.   My
affectionate remembrances to all the family.   God will, I
trust, soon heal their heart wounds.   They have lost a beloved
and honoured father, but he was ripe in years, and he has
left them the inheritance of a good and honoured name.   It
is natural and right they should mourn, but not like those
who are without hope.   God forbid that such sorrow should
be theirs, far otherwise.   Very soon I trust the bitterness of
grief will be over, and they will think of their dear father with
a solemn and chastised joy, and resume the works and
thoughts of their daily life with thankfulness and cheerfulness."

From Mr. Poulett Scrope.

" Let me begin by condoling with you on the loss you have
sustained in common with a large circle of friends and
admirers.

So admirable a person in mind, manner and acquirements, it
will be long before we see again.   The *Mitis Sapientia Lælii*
was never better illustrated, and it was delightful to see so
thorough a disgust for bigotry—in every shape—in Religion,
Science, and Philosophy—coupled with such expansive charity
and benevolence.   A more charming character it never fell to
my lot to associate with."

From the Hon. Edward Twisleton.

" The sad announcement reached me here, and I assure you
how much I sympathize with you in the sorrow which you
cannot but feel.   At your father's age you could scarcely
expect that his life would be long spared to you, but still this
would not make much difference when the blow at length
came.   In addition to his intelligence your father possessed

one of the finest moral natures which can fall to the lot of man, and I shall always hold his memory in honour on this account; but to you he was much more than this, and I can only suggest as a consolation the remembrance of the faithfulness and tenderness with which you invariably discharged your duty towards him."

From Dean Ramsay.

" So old a friend, and a friend so much esteemed. Yes, these are the things which make us sad and thoughtful as time goes on.

" Your father was somewhat older than I am, but he retained his powers and clear intellect in no common degree. I can assure you I heard many expressions of deep feeling when his death was known here. But of course many knew him well, Thomas Erskine, of Linlathen, spoke of him as a schoolfellow and school friend."

From Sir Charles Bunbury.

" I shall always feel it a privilege to have been so long and so intimately connected with so excellent, and so wise a man as Mr. Horner. I shall always feel that I am, or ought to be, the wiser and the better from having known him so intimately, and the recollection of his kindness, and of the affectionate regard that he showed towards me, will always be deeply gratifying.

" I have never myself known a better man, nor one of more universal and unfailing, indefatigable kindness. His active benevolence never cooled in the least the warmth of his domestic attachments, nor did his strong affection for his family and friends check his enlightened ardour for the general good of mankind."

From the Delegates of the Operative Cotton Spinners of Lancashire and adjoining districts in general meeting assembled, to the MISS HORNERS.

Wheat Sheaf Inn, Swan St., Manchester,
19th March, 1864.

HONOURED LADIES,

We are delegates from the Operative Cotton Spinners of Lancashire and adjoining districts in general meeting assembled, who feeling that there are circumstances in life, in which the most enlightened minds may derive some consolation from the sympathy of their fellow creatures, however humble, venture to address to you a few observations on this mournful occasion, and as nothing but the desire to alleviate sorrows which are unavoidable, has dictated this communication, we trust you will accept it, as some apology for this our intrusion upon your privacy.

Actuated by this feeling, we in common with the thousands represented at this Meeting, beg most respectfully to tender to you the expression of our feeling of profound sympathy in the irreparable loss you have sustained in the demise of your worthy and affectionate father ; we most sincerely condole with you on this melancholy event, which in the natural course of things, though not entirely unexpected, is nevertheless one of deepest bereavement to you, and to us of heartfelt sorrow and regret.

It would indeed be a source of the most heartfelt satisfaction to us; could we in any way contribute to mitigate the sorrow which this sad event has occasioned you and yours, or dispel the feelings of melancholy which will in some degree overshadow your future life.

In the hope of affording you some consolation, we beg to be permitted to observe, that your lamented parent has not been

called away early in life, but that it has pleased the Almighty
Disposer of events long to spare him to his fellow creatures,
who have been greatly benefited by his unremitting labours
in the cause of Justice and Humanity. His impartiality in
the administration of the laws made for the protection of
our wives and children, and his firmness in their vindication,
have long commanded our esteem, and of which while we
live we shall cherish a grateful remembrance.

After a life spent in doing good, the mortal remnant of
your venerable parent has descended to the tomb, honoured
by the wise and good of all classes ; while the memory of his
eminent services in their cause is embalmed in the grateful
recollection of the poor factory operatives of these districts.

May these considerations speedily reconcile your minds to
this visitation of Divine Providence by which your worthy
Father and our highly esteemed Friend has been removed
from this earthly scene of suffering and sorrow, affording as
they do the blessed assurance that he has been called hence
to receive the reward of a well-spent life in another and a
better world. May they operate as a balm to your wounded
spirits, and afford consolation to every member of your
respected family. That Almighty God in His goodness may
long preserve and strengthen you all, in this your hour of
trial, and enable you to bear it with patience and christian
resignation, is the most heartfelt prayer of your humble and
sincere friends the delegates,

WILLIAM LEIGH, (*Chairman.*)
THOMAS MAUDSLEY, (*Secretary.*)

INDEX.

THE END.

Women's Printing Society. Ltd., Great College Street, Westminster.

Printed in the United States
By Bookmasters